Networking the World, 1794–2000

Networking the World, 1794–2000

Armand Mattelart

⁂

Translated by Liz Carey-Libbrecht
and James A. Cohen

University of Minnesota Press
Minneapolis London

Published by the University of Minnesota Press
111 Third Avenue South, Suite 290
Minneapolis, MN 55401-2520
http://www.upress.umn.edu

Printed in the United States of America on acid-free paper

Library of Congress Cataloging-in-Publication Data

Mattelart, Armand.
[Mondialisation de la communication. English]
Networking the world, 1794–2000 : Armand Mattelart ; translated by Liz Carey-Libbrecht and James A. Cohen.
p. cm.
Includes bibliographical references.
ISBN 0-8166-3287-1 — ISBN 0-8166-3288-x (pbk.)
1. Telecommunication—History. 2. Technological innovations—History. I. Title.
HE7631.M37513 2000
384'.0921—dc21 99-046654

11 10 09 08 07 06 05 04 03 02 01 00 10 9 8 7 6 5 4 3 2 1

Contents

Preface

Real-time communication networks are shaping the way our planet is organized. The process known as globalization goes hand in hand with fluidity of immaterial exchanges and immaterial flows across borders. This short book proposes to set this new phase of the opening up of the world within the history of the social forms that the internationalization process has assumed over time. The widespread interconnection of economies and societies is the ultimate goal of global integration initiated at the turn of the nineteenth century. By gradually extending the movement of persons and of material and symbolic goods, communication technology has hastened the incorporation of distinctive societies into increasingly larger groups, constantly pushing back physical, intellectual, and mental borders.

Numerous public and private actors have played a part in shaping the topography of networks and systems on a global scale. They have done so in the name of a wide variety of ideals and have been motivated by a broad range of interests: the universalism of a predestined civilization, the ecumenism of a religion, the interdependency of nations based on mutual security, the pragmatism of the corporation, or the categorical imperatives of the international division of labor, or the community of struggle against oppression. Networks, a leading symbol of progress, have also made their incursion into utopian thinking. The communication network is an eternal promise symbolizing a world that is better because it is united. From road and rail to information highways, this belief has been revived with each technological generation, yet networks have never ceased to be at the center of struggles for control of the world.

The homogenization of societies is inherent in the unification of the economic sphere; their fragmentation is the corollary. The gap is widening between market rationality and cultures, between a technoscientific system that is being generalized and the wish to affirm a sense of belonging. This distortion is turning the outcome of humanity's march toward integration into an enigma.

Charting the genealogy of the international sphere is proving to be all the more crucial since the new globalization label tends to be intoxicating. As a matter of method, it is advisable to treat this notion with a measure of doubt and to refute the idea of the social a-topia of words that name the world, and rather to identify the standpoint from which those who conceive and use them are talking. For this term had traveled round the world even before its definition was established as an analytical tool. Its vertiginous rise in importance, underscored by stock market and environmental crises, military expeditions, major sporting events, or other such happenings broadcast worldwide by satellite, is only one side of the coin; the other is the sinking of history into oblivion. The risk is therefore great that it will become increasingly difficult to distinguish dreams from reality in a field of representations already wide open to myths.

Chapter 1
Networks of Universalization

The internationalization of communication was spawned by two forms of universalism: the Enlightenment and liberalism. Two projects, at times opposed and at others convergent, both aimed at the construction of an unrestricted global arena, were trying to materialize: on the one hand the great democratic republics of the revolutionary utopia; on the other, the universal mercantile republic of classical economics.

International communication emerged with modern nationalism, which established the territory as the basis of sovereignty and of an imaginary community. In the wake of the French Revolution, the nation-state, that specific form of organization contained in embryo in the Peace of Westphalia (1648), proliferated. The treaty had terminated the order of the papacy and the Holy Roman Empire and had very symbolically dislodged Latin from its hegemonic position and replaced it with French. In the nineteenth century a system of relationships was established, linking these new national entities on the basis of a rule of law accepted by all. Finally, external borders and internal administrative divisions became less fickle and configured the natural setting for the industrial revolution. Thus was established a productive order firmly embedded in the construction of the international arena and characterized by the international division of labor. In 1802 the French language endorsed the project of this new worldwide political economy by incorporating into its vocabulary the Anglicism "international," a term coined by Jeremy Bentham around 1780.

The Torrent of Enlightenment

Freeing the Flow

The invention of communication as an ideal occurred at a time when the prevailing ideas were those of modernity and the perfectibility of human societies. It was thus the product of a belief in the future. The Enlightenment thinkers prepared for its emergence by advocating trade as a creator of values.

Engineers from the *Ponts et Chaussées* of the ancien régime were among the first to formalize a theory of communication associated with the organization of a national space and the construction of a domestic market, by applying it to roads and canals. They hoped, through such engineering projects, to accomplish the intentions of reason. Through the domestication of a bad nature—the irrational, which separates people and prevents them from helping one another—they believed they could ensure the triumph of good nature—the rational, which brings together, unites, and promotes the free flow of people and goods.

Established as a principle of human rights, the free communication of thoughts and opinions defied borders. Toward 1763 Denis Diderot (1713–1784) sharply addressed the censor in his "Letter on the Book Trade": "Sir, line your borders with soldiers, arm them with bayonets to keep out all the dangerous books which may appear, and these books, excuse the expression, will pass between their legs and fly over their heads and reach us."

The paradox of the Enlightenment was that the authors of the *Encyclopédie,* in order to illustrate the advantages of the free circulation of ideas and goods, referred favorably to China under enlightened despotism. Voltaire (1694–1778) acknowledged the role played there by the gazettes, published by the Peking Court, in the management of public affairs. François Quesnay (1694–1774), the first theoretician of the flow of wealth, referred to the perfection of roads and canals in the Celestial Empire to legitimize the motto of the physiocratic school: *Laissez faire, laissez passer.*

Revolutionizing Language

In 1789 France wanted to implement the idea of the creative power of exchange. By universalizing juridical relations and the circulation of money, goods, and persons, it built up its unity and national identity while simultaneously creating a universal identity.

"One nation, one law, one language." The revolutionary policy of unifying the language was aimed primarily at absorbing differences and breaking the barriers of specific local characteristics inherited from the times of feudalism and absolute monarchy. It intended to do so by eliminating the linguistic barrier separating those who, by virtue of their status, were the only ones likely to understand what was being said from the remaining mass considered incapable of communicating among themselves. The despot, declared Bertrand Barère in 1794 before the Public Safety Committee, "needed to isolate peoples, separate countries, divide interests, hinder communication, stop the simultaneity of thoughts and the identity of movements." Linguistic unity freed the forces of the "torrent of the Enlightenment" and converted each printer into a "public teacher of language and legislation."

In order to free exchange, it was necessary not only to "wipe out patois and universalize the use of the French language" (title of a report by abbot Grégoire in June 1794), but also to "revolutionize the language itself." Condorcet had long since devised a "universal language" that, he claimed, would reproduce the certainty of geometry and be the product of "the application of mathematical methods to new objects." The linguistic policy of the revolutionaries was influenced by a model of universal language: sign language. Many of them were fascinated by the language of the deaf-mute invented by abbot de l'Épée and perfected by abbot Sicard.

Communicating with Signs

The invention of the semaphore telegraph by the Chappe brothers was part of this quest for a "language of signs." The system was based on a linguistic principle: the more signs one has, the fewer one needs to transmit information and the faster this transmission can be carried out. With the inauguration of the first line in 1794, Barère exclaimed: "It is a means which tends to consolidate the unity of the Republic through the close and rapid contact it allows all parties. Modern peoples, through printing, gunpowder, the compass, and telegraphic sign language, have removed the major obstacles to the civilization of mankind." Speculation on the possible civil uses of this technology was rife. Revolutionary thinkers presumed it would suffice to multiply the number of lines and free their coded language, thus enabling all people in France to "communicate their information and their wishes." The conditions of the Greek agora would thus be reproduced nationwide and Jean-Jacques Rousseau's objection against the possibility of "large democratic republics" would lose its validity. Hence, at a very early stage, long-distance communication technology was promoted as a guarantee of the revival of democracy.

We know what became of it. The exceptional rules that had assigned to the telegraph a military function and decreed an embargo on codes became the norm. The language of telegraphic signs was to remain a state secret for a long time. Its use by the public was first authorized over fifteen years after the invention of the electric telegraph in 1837. Utopian socialism compensated for this disregard for citizens' expression by endowing communication technologies with an essential role in the building of the future human community. Anticipating the establishment of the electric telegraph, Charles Fourier (1772–1837) made sign language the basis of universal unity and invented "mirage transmission" to link London to India in less than four hours . . . via the planet Mercury!

Built in the shape of a star around the capital, like all the major networks that were to follow it and the road network that had preceded it, the semaphore telegraph served to communicate with fortresses on the borders and the coast and with other towns. Under the Empire it stretched as far as Venice, Amsterdam, and Mainz, and lost its international dimension only after the fall of Napoleon.

Standardization

Communicating means standardizing and doing away with chance. Although the use of the semaphore telegraph for security purposes, prescribed by the French state, was not in keeping with the dynamic of exchange advocated by the communication ideal, other factors helped to promote this dynamic. In so doing, they also helped to build a framework of interchangeable standards that made universal communication possible. That was the function, in particular, of the standardization of measures, which can be considered as the mortar of commercial transactions and the cadastral basis of the tax system. The adoption of a metric system of weights and measures put an end to the multiplicity of standards linked to local habits and customs. The new stable unit became part of the symbolism of communication. At the National Convention in 1792, the astronomer Joseph de Lalande presented the meter as a "new bond of general fraternity for the peoples who adopt it." Abbot Grégoire went further and acclaimed it as a symbol of national unity and a "beneficial truth that will become a new link between nations and one of the most useful conquests of equality." In 1875, and despite the opposition of the British empire, the meter became an international reference. The universalization of the metric system led to that of the decimal division of money, decreed in 1795 and exported to the United States shortly afterward.

The Outposts of Free Trade

International Division of Labor

In the view of the founder of classical economics, Adam Smith (1723–1790), the merchant was a citizen of the world. The cosmopolitan market was expected to abolish hostile forces that brought nations into conflict and to rid the world of old military societies. As the guarantor of harmony between peoples and nations, the "universal mercantile republic" had the vocation of bringing together the entire human species in an economic community composed of consumers. These consumers were offered goods by producers who, in turn, were supplied by those competitors from whom they could obtain the greatest quantity and quality of the goods required, at the lowest price. In 1776, in his work *The Causes of the Wealth of Nations,* the Scottish economist defined individualism and free competition in a single market governed by the international division of labor as the touchstone of his universal system. Communication was functional to the division of labor; it was as indispensable in the efficient running of the separation of jobs within a plant as it was in the organization of the global workshop.

"To produce is to move": this maxim, so dear to John Stuart Mill (1806–1873), defined the communicational perspective of liberal political economics in the second half of the nineteenth century. The rule of free trade was to govern information and transport alike, for the principle of the free flow of information was a corollary of that of the free flow of goods and labor—a point that Mill expanded on in his *Principles of Political Economy* (1848). He denounced the taxes that hindered the flow of information by burdening publicity, newspapers, and the postal services.

By struggling for the abolition of these "taxes on knowledge," seen as an obstacle to a free press, the political philosophy of liberalism proved itself to be a truly liberating ideal. It argued for the secularization of society, considered individual freedom as the mainspring of institutions, and strove to limit the arbitrary power of the state. With free trade taken as an article of faith, economic liberalism overshadowed these three components and established the de facto determinism of the market. It spawned a market mentality, to use the term coined by the economic historian Karl Polanyi, and a new society, since market mechanisms were spreading throughout the social body.

The deployment of technological networks during the second half of the nineteenth century perpetuated the worldwide economic integration initiated at the turn of the seventeenth century with the expansion of the Dutch East India Company in 1602 and its monopoly of the spice trade. In a world that still seemed to offer limitless possibilities for exploration and exploitation, networks played their part in the new parceling out of the planet that redefined the parameters of national economies.

The First Unified Area of Circulation

The electric telegraph removed the obstacle of national security, which in France had burdened the semaphore telegraph, and triggered a series of bilateral transmission agreements. In the late 1840s a first treaty, concerning the Berlin-Vienna telegraph link, was signed between Prussia and Austria. It was soon followed by a regional grouping, the Austro-German Telegraph Union, and an agreement between Belgium, France, and Prussia. In this domain, as in that of the railways and the postal services, the future Germany proved to be a pioneer in network unification projects. Unification through communication technologies preceded political union in this mosaic of territories.

In 1865 an original institution, the International Telegraph Union, was created to facilitate cross-border telegraphic communication. The function of this organization contrasted with prevailing modes of consultation between sovereign states. Unlike the diplomatic congresses inaugurated in 1815 (which supposedly constituted the embryo of a regular, multilateral political system, yet merely reflected the control that the major powers of the European Concert maintained over international relations) the International Telegraph Union was open to the entire community of sovereign nations. Intended to solve problems in ways that individual states within their borders could not, it prefigured the modern international organization. Its mission was to determine procedures, standards, and common rates between member countries, and to record telegraph traffic. Unlike other forms of intergovernmental agreement, the decisions made were combined with guarantees. The executive function was fulfilled by the secretariat or the international committee under the responsibility not of professional diplomats but of experts and engineers.

The idea of this type of committee was soon adopted by the General Postal Union (1874) (renamed the Universal Postal Union four years later), the International Commission for Weights and Measures (1875) (which marked the success of the metric system), the Agreement for International Regulation of Sea Routes (1879), the International Union for the Protection of Industrial Property (1883), the International Union for the Protection of Literary and Artistic Works (1886), and the Agreement for International Rail Transport (1890). It was applied in widely diverse areas of social and economic life at a time when projects for harmonization were flourishing. According to the German historian Werner Sombart, 17 intergovernmental cooperative agreements of this type were signed between 1850 and 1870, 20 between 1870 and 1880, 31 between 1880 and 1890, 61 in the last decade of the century, and 108 in the first decade of the twentieth century. In parallel with harmonization, which facilitated international exchange, standardization was introduced in the 1880s within the most advanced firms seeking to work with interchangeable parts. The huge demand generated by World War I, coupled with the severe shortage of skilled labor, was to stimulate this process.

Toward 1870 the annual number of telegraph transmissions topped the thirty million mark. By the turn of the century, this figure had increased more than tenfold, and cross-border traffic accounted for one-fifth of these transmissions. The telegraph had already profoundly modified the economic status of information and the methods of collecting, processing, and coding it. It forced speculators to find new ways of intervening in markets and incorporated the most distant parts of Europe into networks of economic exchange. On the eve of the First World War, writes Sombart, the quotations at the Berlin wheat exchange were displayed daily in Siberian villages.

The liberalization of international telegraphic lines created the first unified electric sphere. This initiative contrasted with the protectionist attitude shown by the same contracting states when faced with removing all hindrances to the circulation of goods and with applying literally the liberal trade treaties inspired by the doctrine of free trade that had been adopted by England in the 1840s and by its European rivals two decades later. Although convergence was the order of the day in the field of telegraph networks, this was far from the case with economic integration. The same year in which the Telegraph Union was founded, France signed an agreement for monetary union, the Latin Union, with Belgium, Switzerland, and Italy, using as its currency the Germinal Franc. In 1867, on the fringes of the Universal Exposition in Paris, the government called an international conference and suggested (in vain of course) that the entire world adhere to this single currency.

The Train, a Symbol of the Industrial Nation-State

The first railway line worthy of the name, because it was functional and viable, appeared in England in 1830. The construction of lines on the European continent reached a peak in the 1870s.

The train was above all an emblem of progress and industrial revolution within the framework of the nation-state. More than half a century separated the inauguration of the first line and the creation of the International Railway Conference. As for the standard rail gauge, most European countries (except Spain and Russia for reasons of national defense) adopted that of Stephenson, the British inventor of the locomotive.

Rail gauge standards broke down not only in the colonies but also in dependent sovereign states, for each country and each manufacturer adhered to a different standard. Designed exclusively to meet the needs of the colonial powers, the building of railways was based on the penetration model, to serve the demands of trade and the exploitation of natural resources. This territorialization took on an exaggerated form not only in the African colonies, where the railroad was introduced in the latter quarter of the century, but also in large, politically independent yet economically subordinate states. This was the case in Brazil where, as in the countries of the Southern Cone, the presence of British and to a lesser extent French interests was decisive. At the end of the century there were no less than five autonomous independent local networks, each fanning out from a port and extending toward the hinterland with its mines and plantations. This outward-facing construction also prevailed in the geopolitical area called the "American Mediterranean" (the Caribbean and Central America). One-sided contracts for the concession of telegraph lines, maritime transport, and railways, wrested from local oligarchies in the 1880s by large plantations such as United Fruit Company—the forerunners of the modern agribusiness groups—were closely related to the creation of the notion of a banana republic. The railway imbroglio reached its peak in the Chinese empire at the end of the century when the lines connecting harbors and concessions were built according to no less than five different standards: Russian, Japanese, Anglo-American, German, and Franco-Belgian.

The Worldwide Time of Managers

The organization of the railways was the prelude to the establishment of the universal measurement of time, since the regulation of traffic necessitated the adoption of a national time to replace the multiple local times. The British railways aligned their official time with that of the Greenwich meridian. In 1884, when the international community decided to synchronize the diverse national times, Greenwich time was chosen to serve as a benchmark for the calculation of universal time. This decision was greeted with displeasure in countries such as France (which proposed the time of the meridian running through the Paris Observatory), Spain, and Brazil, which interpreted the measure as a symbolic endorsement of Victorian power and refused to comply with it until 1911.

The railway enterprise contributed two other basic components to the construction of this rationality of modern capitalism with global ambitions. First, the railway companies—and also, to an extent, the telegraph companies—were the earliest modern corporations. They were the first to feel the need to find new modes of organization adapted to the management of the continuous circulation of goods, services, and information on a large scale. Thus, they became a testing ground for modern management techniques. In fact, they invented what business historian Alfred Chandler called managerial capitalism. Chandler saw in them the birth of multidivisional companies, the first to employ large numbers of executive staff to coordinate, supervise, and evaluate the activities of several distinct operational units. Second, the construction of railways throughout the world, relying on large foreign loans, stimulated the internationalization of financial markets and resulted in the steering of corporations by financial capital. Half of the capital exported during the nineteenth century served to finance railways, harbors, canals, and other public infrastructure. Mines, plantations, and industrial enterprises received barely a third, and the rest went to commercial, banking, and other establishments. Commercial banks relegated the major investment banks to a position of secondary importance as they developed their networks of agencies throughout the world.

The Building of World Power

The Undersea Cable and the **Pax Britannica**

The nineteenth century saw the establishment of the British empire as the new economic and financial pole toward which the main flows of wealth and long-distance communication converged. London became the center of a world-economy in the sense given by the historian Fernand Braudel, i.e., a center from which the other powers, intermediate zones, and peripheral regions were organized and in relation to which they situated themselves hierarchically. The undersea cable was one of the clearest illustrations of Victorian hegemony.

The first undersea cable was inaugurated in 1851. It connected Calais to Dover and Paris to the City of London. Fifteen years later, after three unsuccessful attempts, the first transatlantic cable was laid. Then a line was laid between Malta and Alexandria, which enabled London to communicate directly with India. In the 1870s the British network spread toward Southeast Asia, Australia, and China, and also toward the West Indies and South America. Africa received cable a little later, in the late 1880s and early 1890s. The last link in the British global network, the transpacific, was completed in 1902. At that point the cable system of the Victorian empire accounted for two-thirds of the world network and its fleet of cabling ships was ten times larger than that of France. So tightly entwined were the logics of trade and diplomacy that the fact that the British undersea cable depended on private companies—unlike France, where it was placed under state control—changed nothing from a geostrategic point of view. Supported by the Admiralty and the know-how of its cartographic services, the expansion of private companies relied on the control of raw materials (copper and rubber), on financial power, and on the mastery of cable technology (production, installation, and operation). This, in turn, enhanced the supremacy of the navy and the great London-based shipping lines. Other governments were obliged to use the services of the British system for their own communication. Thus, in 1898 with the outbreak of the Fashoda crisis, in which French westward colonial expansion in Africa was confronted by the north-south expansion of the British, Paris was able to communicate with Sudan and the leader of the French expedition only through the networks controlled by its rival.

The first signs of opposition to British hegemony appeared in the 1890s. Because the British Post Office had refused to allow a German transatlantic cable to cross its territory, Berlin decided to make an all-out effort to master the technology at all levels, from production through operation. In 1900 the Kaiser inaugurated the first Emden–New York cable via the Azores, followed by another two years later. In 1903 the United States laid its own transpacific cable linking San Francisco to Manila via Honolulu and Guam. Five years earlier it had snatched the Philippines from a moribund Spanish empire.

Sharing the Airwaves

In 1901 Guglielmo Marconi inaugurated the era of transatlantic radio communication when his transmission of the letter *s*, from Britain, was received on the other side of the ocean. The Navy, the War Office, and the Post Office were the first to show an interest in the Italian engineer's patents. The acquisition of exclusive property rights by the British firm Marconi Co. marked the start of the Victorian attempt to create an institutional framework for the internationalization of this new technology, inspired by the model on which the undersea cable had been based. However, Germany, which had patents on technologies developed by the firms Siemens and AEG, and which founded Telefunken in 1903 to put them to use, intervened. At Germany's initiative, two international conferences were held in Berlin in 1903 and 1906. The English argument for the need to impose a single type of apparatus for the transmission of signals was outvoted, and the free competition opened within the strictly private club of owners of technological patents was restricted to military use until the end of World War I. The 1906 conference marked the founding of the International Radiotelegraphic Union, which dealt with the problem of interference and laid the foundations of an unequal order of world communication. The maritime powers, the main users of these technologies, imposed the rule of "first come, first served." In order to use a wavelength, a country had merely to notify the Union of its intention to do so. This rule led to the virtual monopolization by a minority of the world's radiophonic spectrum. It expressed, in the communications field, the increasingly unequal exchange between productive systems, a situation that developed in tandem with scientific and technological progress. In 1800 the difference between the gross domestic product per capita in the

North and the South was insignificant; by the end of the Age of Empires (1875–1914), the ratio was three to one.

From the start of the twentieth century, socialists of all leanings sought to define the logic of exchange value characterizing unequal development. In this context the critical version of the concept of imperialism was born. At the heart of their analyses lay the formation of large cartels and trusts for the purpose of organizing the market, that is to say, limiting the competition that Adam Smith wanted to be free—in particular, the cartels of the electrotechnical industry, the energy trusts, and the railway companies, in partnership with the iron and steel industry.

During the Berlin Conference of 1906, the international regulation of the telephone, an invention patented by Edison in 1876 and whose use was governed until then by bilateral agreements, seemed to represent smaller stakes than that of radio communication. Admittedly, the internationalization of this network was still limited. The first telephone transmission from Paris to Brussels occurred in 1887, and from Paris to London three years later. The telephone did not acquire a truly international dimension until September 1956, with the inauguration of the first transatlantic telephone cable, just a year before the launching of the first artificial satellite.

War and Geopolitics

A close relationship can be observed between the rapid growth of communication technologies and the armed conflicts that broke out during the latter half of the nineteenth century. During the Crimean War (1853–1856) the first cable was laid across the Black Sea and direct telegraph lines were established in the field and between the army headquarters and the governments of London and Paris. The invasion of Mexico in 1846 by its northern neighbor, marking the first explicit formulation of the expansionist doctrine of Manifest Destiny, revealed the usefulness of the electric telegraph in military operations and in the transmission of news. The American Civil War (1861–1865) taught the military strategists their most valuable lessons on the uses of the "iron horse" and telegraph lines. Many armies, starting with the Prussians, drew on this experience in devising their logistics (the art of moving armies) and added communication troops to their forces. Before the First World War, the Anglo-Boer War (1899–1902) and the Russo-Japanese War (1904–1905) confirmed, respectively, the decisive impact of the train and the telegraph, and of radio communication. Britain was quick to learn from this experience by making radiotelegraphy a state monopoly and incorporating it in the Post Office, over which the Admiralty kept a close watch.

Geocommunicational complexes could not be controlled, however, without provoking national and international tensions. In China the movement for the buyout and nationalization of the railways demanded the construction of lines toward the provincial capitals and instigated the republican insurrection of Wuhan and the downfall in 1911 of a thousand-year-old empire. In the context of the "Eastern Question," the Ottoman Empire, by granting concessions to the German empire in 1911 for a cable between Constance and Constantinople and for a railway line toward Baghdad, to be extended toward the Persian Gulf, provoked a state of ferment between rival empires. Britain and France saw in it the expression of a pan-Germanic expansionist policy aimed at implementing its motto *Drang nach Osten* (the rush toward the East) by consolidating its position in a region that opened up onto the oil fields. Short-circuiting the British empire by bypassing the Suez Canal was a constant obsession in expansionist rail strategies toward the East. It underlay the pan-Germanic project and presided over the construction of the trans-Siberian railway. The Moscow-Vladivostok line (8,156 km), started in 1891 and completed in 1903, opened the window onto the Far East and afforded the czar's empire excellent strategic positions.

In the context of the pan-Germanic movement, Friedrich Ratzel published the first geopolitical treatise in 1897, entitled *Politische Geographie,* that laid the foundations for a science of space, the forerunner of a science of networks. Networks were recognized as vitalizing a territory. The author had grasped their importance through his empirical analysis of the fully dynamic U.S. territory. In his theoretical text he forged the concept of world power and addressed the spatial dimension of international relations from a global point of view. We also note in it the emergence of an ideology based on the metaphor of living organisms, the so-called spatialist ideology, with its notions of living space (Lebensraum) or of natural boundaries, which was to become a source of legitimacy for many future ventures in expansionism. Living space was seen as an expression of animal territorial laws that justified war, conquest, and encroachment.

The Utopias of Universal Communication

Universal Association

In the first half of the nineteenth century, the basis of an ideology of redemption through international communication was consolidated in France. "Embrace the universe" and "Everything through steam and electricity" were the watchwords of the disciples of the French philosopher Claude-Henri de Saint-Simon (1760–1825).

At the turn of the 1820s, in contrast with the economistic vision of the international division of labor proposed by Smith, accused of widening the gap between rich and poor, Saint-Simon proposed the utopias of the "universal association through industry," the exploitation of the world by "men in partnership" working, with a common motivation, toward the accomplishment of a common goal. The earth, thought Saint-Simon, must be "administered" by industrialists as a "great industrial corporation" rather than "governed" by a tutelary state. This axiom was the basis of "positive knowledge" about the management of human beings. In this project of planetary restructuring, the network, as a model of rationality, became the emblematic figure of the new organization of society.

Saint-Simon saw this theory of social reorganization as enabling contemporary society to emerge from its dual crisis. Part of the crisis dated back to 1789 and was rooted in the "negative knowledge" of the Enlightenment thinkers and its revolutionary excesses. Although legitimate when it was a matter of undermining the earlier order, this critical attitude became counterproductive in the creation of a new social order and the "shift from the feudal and theological to the industrial and scientific system." The crisis also stemmed from the fact that Europe was "disorganized" and unable to reconstitute the "system of international peace" that had been lost since the end of a unified Christian world.

The Determinism of Networks

Saint-Simon remained a man of road networks, an admirer of the *Ponts et Chaussées,* the public body responsible for the construction and administration of roads and bridges. In 1832, seven years after his death, his follower Michel Chevalier (1806–1879), one of the founders of the ephemeral Saint-Simonian church, adopted a deterministic conception of the networks of what he called "circulating civilization." Yet the railway and the locomotive had by no means revealed their full potential for structuring spaces. England had built the first railway line worthy of that name only two years earlier, while the French authorities were still busy calculating the merits of the invention. Only in 1842 was a law passed to authorize the founding of a national rail network. Chevalier nevertheless prophesied society's future.

Like his teacher, he considered that "spiritual" or credit networks and "material" or communication networks had a function of cohesion in the social organism. Railways, in conjunction with shipping lines and long-distance communication, were, in his view, the vehicles of Universal Association. This association was to start with the formation of a "Mediterranean System." Its engineers and workers would come from the army, which would be reoriented to civil service. Communication was destined to replace religion (from the Latin *religare,* to link) because, like religion, it had the function of linking disconnected members of an underlying community and drawing numbed civilizations from their torpor, from Greece to Asia Minor and from Spain to Russia. The solution consisted of "placing close to them examples of an extraordinary movement, exciting them with the show of a prodigious swiftness, and inviting them to follow the current that flows past their door." The question of democracy, which Chevalier treated as a dependent variable, subordinate to technological and industrial development, was far from the center of his concern to organize the world. This did not, however, prevent him from repeating that communication reduces distances not only between two points but also between social classes. Thus, improving communications necessarily promotes equality and democracy.

Once the Saint-Simonian church had been dissolved and the grand visions of the militant period had ended, Saint-Simonism was content to express a managerial thinking before its time and symbolized the spirit of enterprise of the second half of the nineteenth century. The ideology of redemption through networks, seen as creators of a universal bond, legitimized managerial positivism. The new entrepreneurs of industrialism laid the foundations of international networked space by creating railway companies and shipping lines, founding credit institutions, and building canals between oceans.

The International Dimension of Social Networks

In keeping with the aspiration for a more just society originally conceived by Saint-Simon, those who were disillusioned with Saint-Simonism turned away from the technicist vision of networks presented as the key determining factor of a new society. Many became precursors of socialism, placing all their hopes for "embracing the universe" in a future socialist society. Many could identify with Pierre-Joseph Proudhon (1809–1864), in his book *Des réformes à opérer dans l'exploitation des chemins de fer* (Reforms to be adopted in the operation of railroads, 1855), in which he criticized those who mistook the "mercantile movement" for the "intellectual movement": "Ideas are circulated not by coaches but by writers, political discussion, the free press. . . . The length of railway lines in operation in France has tripled. Since then we have not seen the slightest idea circulate."

In the context of thinking focused on the social network, the notion of "internationality" appeared in the writings of a feminist pioneer, Flora Tristan (1803–1844), in 1843. Tristan herself was at a cultural crossroads, born of a French mother and a Peruvian father. The basis of her project for a Workers' Union was the new principle of internationalism (an idea that appeared five years later in the *Communist Manifesto* of Marx and Engels). In the table of contents of the Workers' Union weekly, the first title mentioned was "General interests, that is, international European and worldwide interests." "Democratic cosmopolitanism" became the rallying cry of numerous movements that, matching action and words, created their own press, often had their own singers and poets, and traveled the roads to spread their ideas of "fraternity and solidarity between nations and individuals." The First International of workers was institutionalized in London in 1864. In its statutes we read:

> Since emancipation from work is neither a local nor a national problem but a social one, it encompasses all countries in which a modern society exists. Its solution necessitates the theoretical and practical cooperation of the most advanced countries. . . . The Association is established to create a central point of communication and cooperation between the worker societies in the different countries which have the same aspirations, that is to say: mutual assistance, progress, and total freedom of the working classes.

Ten years later the First International was dissolved. Universalistic thinking was in crisis after a triple failure: the Franco-Prussian war had stimulated patriotism; the crushing of the Paris Commune had sounded the defeat of the minority of internationalists and their motto "The flag of the Commune is that of the universal Republic"; and, lastly, differing conceptions of the state apparatus and its aims and means of action had divided the workers' movement. Soon these differences were to determine the orientations and international alliances of the major trade unions. Before its dissolution the First International was the locus of opposition between two conceptions of the state and of the idea of public service as applied to the management of railway networks.

Between 1830 and 1875 the novelist Eugène Sue, a sympathizer of early Saint-Simonism, used his serialized novels to spread his utopia of social reform and peaceful reconciliation between rich and poor. The stories of this former navy surgeon, profoundly influenced from the start by the English Roman-Gothic school and its main exponent, Ann Radcliffe, popularized a new genre that was the precursor of mass culture and the culture industry and one of the first expressions of serialized literature to cross international borders. As for Jules Verne, he perpetuated the generous vision of the first utopian socialists through the exploits of his Promethean heroes of progress. In an opus that appeared between 1860 and 1906, Verne frequently depicted the networks of steam power and electricity that were covering the globe.

The Universal Expositions

The great Universal Expositions were a medium that was favored by Saint-Simonism and that contributed to the development of the imaginary of communication. These grand events helped Paris, which organized five of them, to become the "cultural capital of the nineteenth century," in Walter Benjamin's phrase. Within these "peaceful microcosms of progress," the products of the economies of various nations were exhibited, while a large number of international, governmental, and nongovernmental associations were founded and discussed the most diverse subjects in their congresses. Exhibitions and technological inventions reinforced one another in the propagation of a rhetoric of peace and communion between peoples where "All men become brothers." Each technological generation provided a new opportunity to propagate the grand narratives of general concord and social reconciliation under the aegis of Western civilization. Very symbolically, the first Exposition, held in the Crystal Palace in London in 1851, took place just as the first international undersea cable between Dover and Calais was inaugurated. The one that closed the century, in 1900 in Paris, witnessed the triumph of the cinema.

Cinema was to open the way for the mythology of universal communication in the "era of images," which became another symbol of the end of inequality between classes, groups, and nations. The American novelist Jack London (1876–1916) wrote in the February 1915 issue of *Paramount Magazine* that animated images were tearing down the barriers of poverty and of the environment that barred the route to education, and distributing knowledge in a language that everyone could comprehend.

> The worker with a poor vocabulary is equal to the scientist. . . . Universal education is the message. . . . Time and distance have been annihilated by the magic film to bring together the peoples of the world. . . . Gaze horror-struck at war scenes and you become an advocate of peace. . . . By this magic means, the extremes of society take

> a step closer to each other in the inevitable readjustment of the human condition.

A half-century earlier, at the signing of the treaty marking the founding of the International Telegraph Union, the French Minister of Foreign Affairs welcomed delegates from twenty countries in the following, equally messianic, terms:

> We are gathered here in a genuine Congress of peace. If it is true that war, more often than not, is born out of misunderstanding, are we not removing one of its causes by facilitating the exchange of ideas between people and by placing at their disposal this amazing transmission system, this electric wire through which thought can travel across space at the speed of lightning, and which permits swift and uninterrupted dialogue between the scattered members of the human family?

This passage shows, once again, how recurrent was the association between the ideal of peace and the image of the communication network. In 1849, Victor Hugo, with the leader of the English free trade movement, Richard Cobden, at his side, opened the discussions at the International Congress of Peace, held in Paris, with these words: "How peoples touch each other! How distances are growing shorter! And growing closer is the beginning of fraternity. . . . In a short time, man will travel the Earth just as the gods of Homer traveled the sky, in three steps. Just a few more years, and the electric wire of concord will embrace the whole world." The dream of reestablishing the pre-Babel "great human family" is present throughout the history of the imaginary of communication networks.

The New Arcadias of Electricity

Even before its industrial and domestic applications, electric energy fed the imaginary of communication. In 1852 a book by Michael Angelo Garvey, *The Silent Revolution or the Future Effects of Steam and Electricity upon the Condition of Mankind,* envisaged the social harmony of humanity on the basis of a "perfect network of electric filaments." At the end of the nineteenth century the Russian anarchist and geographer Peter Kropotkin (1842–1921) and the Scottish sociologist Patrick Geddes (1864–1932), both severe critics of the plundering effects of industrialism, saw electricity as the starting point of the "age of neo-technics."

In opposition to the liberal notion of the division of labor that erects barriers between groups, classes, peoples, and nations, and to the neo-Darwinian interpretation of history as competition for life, Kropotkin proposed a history of successive forms of "mutual and reciprocal aid" and the progressive integration of human groups. This was the only parameter, in his view, for judging the evolution of the earth toward a world community. Electric energy was the means for returning to this history of community logic that brought human beings into solidarity with one another. By casting off the weight of the "era of paleo-technics" characterized by mechanics, industrial and urban concentrations, and expansionist empires, this new stage in human history would spawn a horizontal and transparent society. Only the model inspired by industrial ideology could hinder the development of this liberating potential of electricity. Through deconcentration and decentralization, the new energy would open an age of reconciliation between town and country, work and leisure, brain and hand. By contributing to debate on regional development, this current of thought was to influence utopian visions of urbanism for a long time to come.

The contrast is striking between utopian discourse on promises for a better world due to technology, and the reality of struggles for control of communication devices and hegemony over norms and systems. In 1881, at the dawn of the neotechnological era, the first International Electricity Show was held in Paris, with Thomas Edison as an honored guest. This event brought together representatives of the powerful patent-holders of this invention to determine the universal units of measurement such as the ampere and the volt. Unlike the Universal Exhibitions, no sovereign peripheral country was invited to this summit of core countries of the industrial world.

Chapter 2
The Culture Factory

The nineteenth century was the age of the invention of news and the ideal of instantaneous information. The major news agencies were founded between 1830 and 1850, and in 1875 the first press group came into being. The earliest written genres in mass culture also appeared during that period. By the outbreak of World War I the film and music industries had already revealed their export potential.

The Information Industry

Agents of News Value

The world news system depends on the collection and dissemination of information by news agencies. The intervention of these agencies involved not only networks of correspondents over the entire globe, but taking part in undersea cable projects.

Havas, the forerunner of Agence France-Presse (AFP), was founded in 1835, the German agency Wolff in 1849, and the British Reuters in 1851. While Havas combined news and advertising, Reuters focused primarily on economic information. The U.S. agencies AP (Associated Press) and UP (United Press) were launched in 1848 and 1907 respectively, but only the three European agencies were international in scope. Through a treaty of alliance signed in 1870, this triad carved up the world into territories or spheres of influence, thus marking the birth of an information market conceptualized on a global scale and based on geopolitical interests. Each party undertook not to distribute in the territories of the other two. Reuters laid claim to the entire British Empire, Holland and its colonies, Australia, the East Indies, and the Far East; Havas obtained France, Italy, Spain, Portugal, the Levant, Indochina, and Latin America; Wolff focused on central and northern Europe (markets that were taken from it at the end of World War I). By common consent some territories, such as the Ottoman Empire and Egypt, were covered by all three agencies, whereas others, such as the United States, were declared neutral. This oligopolistic organization strengthened the monopoly of each agency over its own national market. The law of cartels and territories was to last for over fifty years.

Despite the late arrival of U.S. agencies on the international scene, their paradigm was to have a profound impact on the French press at a crucial time when dailies, emancipated by the 1881 law on freedom of the press, had set their sights on a mass market. In 1883 the English-language newspaper, the *Morning News,* was founded in Paris. *Le Matin,* inspired by this brief experience and run by a group of journalists of whom many were from Britain and the United States, appeared in 1884. Shortly afterward the *New York Herald* launched a European edition from the French capital. For the Parisian dailies and Havas, which still favored political and diplomatic news—considered to be the noble side of the profession—this development came as a shock. They found themselves confronted with a model of journalism that gave priority to news value and human interest; news that was useful, fast, concise—much like a telegram—and blown out of proportion, with journalists who tracked down trivial events. According to the media historian Michael Palmer, the notion of Americanization can be traced back to this initial encounter with the professional model from across the Atlantic. This notion was developing equally rapidly in the entertainment world. In 1889 Buffalo Bill Cody and his gaudy troupe of "Redskins" arrived at the Universal Exposition. The Parisian press dared to compare two styles of leisure: classical French drama by Corneille, performed by actress Sarah Bernhardt, and the riding antics of Cody, known as the Napoleon of the Prairie. The contrast between so-called high and low culture was already becoming apparent.

The year 1898, a turning point in the legitimization of international news, was marked by three events that were rich in "human interest": Fashoda, the Dreyfus affair, and the Marines' landing in Cuba. To precipitate war on the Caribbean island, one of the last remaining possessions of a moribund Spanish empire battling against the native population, the yellow press of William Randolph Hearst—on whom Orson Welles based his immortal *Citizen Kane*—unleashed a massive indoctrination campaign backed up by street demonstrations. Pictures of poverty and famine, pictures of starving women and children packed into concentration camps (the *reconcentrados*) to avoid all contact with the rebels, traveled around the world. This mobilization of emotions became the alibi of imperial intervention of a new, noncolonial kind. A well-known anecdote neatly summarizes the extravagance of the times. Hearst dispatched reporter and renowned cartoonist Frederick Remington to Havana. From the Cuban capital Remington sent a cable to his boss, saying: "Nothing to report. All's calm. There won't be war. Want to go home." Hearst immediately replied: "Please stay. Provide pictures, I'll provide the war." During the First World War such abuses by the yellow press were to serve as a counter-reference to justify wartime censorship by the French army.

The last decades of the century were to prove as decisive in the field of financial news as they were in the relationship between war and information. In 1888, about a hundred years after the launching of the *Times*, the oldest of the modern dailies, the *Financial Times* was published for the first time in London. New York's *Wall Street Journal* appeared in the following year.

Strategic Information

The first systems for observing and analyzing the international market were set up in parallel with press news. In 1899 the archetype of the modern advertising agency, J. Walter Thompson, founded about forty years earlier in the United States, set up a consulting firm in London for European industrialists wanting to export to the United States. In 1888 this agency published the first bilingual (French-English) guide introducing Europeans to the ins and outs of the American market and press. It also created a Latin American department at its headquarters. The first advertising campaigns for American brands, launched in Europe by the London subsidiary, appeared in the 1920s.

Firms specializing in commercial information (which initially consisted mainly of details on the debts and solvency of firms) were established in the 1830s in England, around 1840 in New York, in 1857 in France, and three years later in Germany. On the eve of the First World War, Berlin was the seat of one of the world's largest strategic information firms, the outcome of a merger in 1887 between W. Schimmelpfeng and Bradstreet Company. In 1890 this firm, whose business ranged from systematic files on firms based on published statistics to industrial espionage, employed 106 persons. By 1914 this figure had swelled to 2,400, and the number of subsidiaries had shot up from 15 to 100. This culture of strategic information, based on the methods used by the German army, was imported into the consular corps, where it led to profound restructuring.

The legitimization of economic information in nineteenth-century Europe was the culmination of a long process. It confirms the American historian David Landes's theory that one of the factors explaining the lead of Western societies in the industrialization process was the "passion for learning from others." Innovation rhymed with imitation. The practice of industrial espionage is a common theme running throughout the modern history of Europe, especially since many non-European societies were less enterprising in this field and yet had made significant progress in essential areas of technology (e.g., the crank, the compass, gunpowder, paper, and most probably printing, all imported from China).

Toward the Industrialization of Culture

Early Genres of Mass Culture

Serialized literature or "industrial literature," a term coined at the time by Charles A. Sainte-Beuve, assumed its definitive form in France during the 1830s. The genre lay at the intersection of popular literary traditions from cultures as different as those of England and Spain. Serialized novels, the first exportable form of a culture intended for the masses, became the vehicle of a true "international sentiment." Through abundant translations, the model was adopted in numerous countries and adapted to local mentalities. The development of the genre was closely related to the history of the press, for the press had nurtured it as a way of increasing circulation, much like advertising and the other pioneer genre comics.

While industrialized culture inherited the melodramatic genre from Europe, the United States gave it comic strips. These appeared in the Sunday edition supplements of dailies in the last three decades of the century, during the bitter struggle between the New York press magnates Joseph Pulitzer and William Randolph Hearst. Comics started to develop their own set of conventions around 1895. Division of labor and standardization of production went hand in hand with internationalization, and the latter was guaranteed by a new type of mechanism, the syndicate, an omnipotent intermediary. As sole holder of copyrights, it could alter dialogues and choose different authors. Practices of this kind, which granted copyrights to a single producer, explain the United States government's stubborn refusal to become a member of the Bern International Union for the Protection of Literary and Artistic Works, ratified in 1886 by ten states only. A century later the United States was still disputing the concept of the "moral rights of authors," the cornerstone of the Bern Convention, of which one of the initiators had been Victor Hugo.

The first syndicate, International News Service, was created in 1909 by Hearst. Apart from its comics, the agency sold the press a wide variety of reproduction and translation rights on articles, reports, games, and crosswords. In 1915 the Hearst group formed a second syndicate, King Features, which was to become the world's largest comic distributor. In 1929 the news agency United Press also formed a syndicate, United Feature. The first comic to attain the international market, George McManus's family strip *Bringing Up Father* was the property of King Features. The fact that it was a product intended for a wide family readership and that it displayed the family unit was not a matter of chance. This moral theme, chosen at the expense of other less standardized and conventional products available at the time, appears to be one of the first common denominators in the agglomeration of a heterogeneous international public.

Historians of the comics underscore the fact that well before their birth, another source of picture stories or graphic sequences had been born in Europe. Perpetuating the tradition of caricature that thrived in England in the eighteenth and early nineteenth centuries, Rodolphe Töpffer (1799–1846), a Swiss novelist, playwright, and artist, had inaugurated a new genre in 1820. During the author's lifetime his "engraved stories," originally in French, had been published in several foreign countries. However, with their extremely limited circulation, it was certainly not in this way that this precursor of comics was to serve internationalization. Its interest lay in the questions raised by Töpffer on the typology or characterization of his characters and their individualization through the permanent features differentiating them. The masters of serialized novels during that period faced the same questions. We know that Balzac, in his *Comédie humaine* and Sue, in *Les Mystères de Paris,* were inspired by the models of characters established by means of physiognomy, very much in fashion since the late eighteenth century. This pseudoscience claimed to establish a relationship between the "visible surface" and "the invisible lying beneath it," and thus to identify links between the face and the personality, judging character by facial features and expressions. We also know of the extremes to which these attempts at psychological decoding were taken when criminal anthropology used them in the last quarter of the century to draw up profiles of the typical delinquent. In 1845 Töpffer wrote a critical essay on the use of these standardizing graphic typologies, in which he warned against the risk of schematization and oversimplification.

Sound and Animated Images

In 1877 Edison invented the phonograph, and at the turn of the century the replacement of the cylinder by the 78 record finally launched the record industry. In 1895 the Lumière brothers projected their first film. The phonographic and cinematographic industries were born, and from the outset they had an international scope.

In 1897 Pathé Frères invested in the music industry, and Deutsche Gramophon and the British firm The Gramophone Company were established a year later. The U.S. company Victor Talking Machine was founded in 1901. In the following year the first satisfactory recording, by Enrico Caruso, was put on sale. A record by the same opera singer, recorded in 1904 in Milan, was the first to top the million copy mark. By crossing the Atlantic, Caruso's records provided Italian immigrants with a link to their home country. From the first years of the twentieth century the major phonographic companies possessed an international network of local agents. In 1908 The Gramophone Co., firmly established in the European market, set up a factory in Calcutta and studios in Bombay, which exported to East Africa.

Cinema spread so fast that many Asian and Latin American countries learned of this new technique at the same time as Europe or parts of the United States. Films were shot in countries ranging from Egypt to Mexico, Brazil, China, or India. Yet, even in India, which subsequently gained control of its domestic market and became one of the major film producers worldwide, the progressive construction of an international film market occurred at the expense of local productions.

Pathé, an emblem of international enterprise prior to the First World War, opened subsidiaries in New York, Moscow, and Brussels in 1904 and, in the following six years, in Berlin, Vienna, Saint Petersburg, Amsterdam, Barcelona, Milan, London, Budapest, Istanbul, Calcutta, Warsaw, and Rio de Janeiro. As film producer and distributor, the French company controlled the entire industry, from ownership of cinemas to the manufacturing and sale of cameras and film. Before 1914 the de facto monopoly of the French producers Pathé and Gaumont was such that the only alternative for countries like England and Germany was to focus their efforts entirely on distribution and, in particular, commercial exploitation. With few exceptions, films had to be imported from France. Denmark and Italy, with two relatively successful production companies, trailed far behind. In the United States, after abortive attempts at horizontal and vertical unification, the period before the war was characterized essentially by the founding of Hollywood. Detached from Los Angeles in 1913, the future capital of American cinema was the outcome of a patent war (1909–1914) waged by independent producers who refused to pay for licenses and were consequently forced to move far away from New York and set up close to a border where they would be able to hide their equipment in case of seizure. Like comics, another iconic medium, animated pictures proved to be a powerful instrument for melding together immigrant populations. The First World War drove the American film beyond its national borders.

The Nature of Publics

The advent of a press intended for the broad majorities set the terms for a debate on the emergence of the democracy of opinion. Conservative stereotypes, spawned by opposition to the French Revolution, would sometimes reappear. The blurred memory of the revolution, associated with collective brutality set loose by convulsive crowds, was the basis of a representation of the collective as the mob. Renewed with each uprising, strike, or violent demonstration, it was validated again in the last decade of the nineteenth century by crowd psychology. Related to the assumptions of criminal anthropology, this approach to the collective, which claimed to explain the sudden appearance of the masses in the life of the community in terms of social psychopathology, pervaded the debate on the effects of greater freedom of the press and association. It was from this perspective that the new forms of assembly and gathering appeared as threats to the established order and became synonymous with "cultural regressions," for a crowd could act and react only as a sleepwalker—hypnotized, hallucinating, subject to contagion, impulsive, credulous, and irrational. That was the viewpoint, for example, of Gustave Le Bon (1841–1931) in his *Psychologie des foules* (1895).

Others, however, thought that the crowd was a phenomenon of the past and that the public was one of the future. They saw society as being increasingly divided into publics superimposed on religious, economic, aesthetic, and political divisions, as well as on corporations, sects, schools, and parties. In contact with internationalization, these new groupings were bound to become more complex. For Gabriel Tarde (1843–1904) in *L'opinion et la foule* (1901), journalism was "a pump for propagating information to all ends of the world." According to him, the fact that some major newspapers, such as the *Times* and *Le Figaro,* and certain major journals, already had their readers scattered throughout the entire world pointed to the advent of "essentially and consistently international" publics.

On the eve of World War I, debate on the nature of the public and its corollary, the persuasive impact of the press on its readers, was nevertheless dominated by a so-called diffusionist perspective in which influence always spreads out from a tutelary center that imposes its worldview on the diverse peripheries. In towns, workers took the middle classes as their model; in the country, peasants saw workers as a reference. On the international scene, in order to know what the future would consist of, the "less developed" nations inevitably had to look toward those which had attained a high level of "civilization." The idea of influence in a single direction was consubstantial with the ideology of linear and continuous progress. It was the basis of the dominant notion of civilization.

The Missionary Press

Alongside the popular press, the networks of the Catholic missionary press, an important crossroads of international representations, continued to develop. In this respect France, the "eldest daughter of the Church," was a hub. In 1822 the *Annales de la propagation de la foi* was founded in Lyon, with the blessing of the Roman Catholic congregation. This bimonthly publication was the mouthpiece of a vast system that collected donations and alms to support the Roman Catholic church in the task set for it by the pope after the fall of Napoleon: "to cover the earth in a network of missions." In 1868 the *Annales,* initially created to perpetuate the tradition of publishing missionaries' letters from China and Canada, which the French Jesuits began in the seventeenth century, was transformed into a weekly entitled *Les Missions catho-*

liques. At the time it adapted to current changes in journalism by incorporating more and more concrete information on the "glorious march of evangelism." Translated into many languages, the Lyon-based periodical established itself as a model of catholicity for other nations. At the end of the First World War, over four hundred Catholic missionary periodicals in different languages existed throughout the world.

This vigor of the denominational press contrasted with the official doctrine of the Vatican concerning the right to freedom of expression. The Roman Catholic Church had previously blacklisted the *Encyclopédie*; in the following century it opposed liberal French Catholics' demands for freedom of the press.

Necessary Interdependence

The World as a Giant Insurance Company

The nineteenth century hailed communication as the agent of civilization. Its networks spun a representation of the world as a vast organism with solidarity between all its parts. The biomorphic notion of "interdependence"—based on the image of the interdependence of cells—validated this widespread impression of interconnection between individuals and societies. The term internationalization also became established at the end of the century, first in the English language and then in the Romance languages, which borrowed it from English.

This shared dependence was seen as linking together everyone and everything in space and in time; it was a form of organic solidarity that showed the way toward a new type of social organization guaranteed by universal insurance and the sharing of risks by all. The nation and the entire world became, in a sense, a giant mutual benefit insurance company run by the state, which, by calculating risks and setting the premiums to be paid by each person, functioned on the basis of reciprocity. In the context of the nation-state, this principle, derived from the application of calculated probabilities to the management of public life, was the start of the welfare state and its system of social insurance. In the field of international relations, it paved the way for the doctrine that was grounded in the legitimacy of the first international system of solidarity and calculated reciprocity, the first institution responsible for guaranteeing mutual security: the International Labor Office of the League of Nations.

Uniformization of the Planet: Science-Fiction?

The idea that the interdependence of nations pushed the world inexorably toward cultural unification took off at the turn of the century. Novelist H. G. Wells (1866–1946) inaugurated the debate in his essay *Anticipations,* based on the following question: Which language will prevail in the third millennium in Europe and the world? And with the language—the prime locus for the definition of a national cultural identity—which culture will prevail? The question was indeed a burning one. In reality France, whose language had been the lingua franca of international relations for almost 250 years, had already felt the base of its linguistic predominance cracking under the influence of other tongues. In order to stand up to this "Darwinian struggle" for linguistic hegemony, it created the Alliance Française in 1883, a "national association for the propagation of the French language in the colonies and abroad."

Wells was by no means of the same opinion as the prophets of doom who took for granted the supremacy of the English language. According to him, in the year 2000 two or three languages would "lay claim to the empire of the world," but the main match would be between French and English. French had serious advantages over its direct rival, starting in Europe where the third millennium would begin with the accomplishment of the dream of European confederation, foreseen in the early nineteenth century by philosophers such as Saint-Simon. And whoever reigned over the continent that was the guardian of universal civilization would spread their influence across the universe. French would gain ascendancy because the public influenced by French culture "extends far beyond the borders of its political system." Did French not enjoy the advantage of the high level of its scientific, philosophical, and literary works? Wells thought the situation was very different in English-speaking countries, especially in England, where the kind of books that predominated were "novels adapted to the minds of women or of boys and superannuated businessmen, stories designed to allay rather than stimulate thought—

they are the only books, indeed, that are profitable to publisher and author alike." Unless there was a "cultural renaissance" and a change of attitude in the "small class that monopolizes the management of business, incapable of understanding the political meaning of the question of language," English could not hope to dislodge French from its position. We know what became of that prediction with the industrialization of culture.

Still, according to Wells, all the forces going against the maintenance of local social systems and leading the world toward the adoption of one or two "unifying languages"—which he deduced by extrapolating from the realities of his time (pan-Americanism, pan-Latinism, pan-Germanism, pan-Slavism)—did not necessarily imply homogeneity. For, "the greater the social organism, the more complex and varied its parts, the more intricate and varied the interplay of culture and breed and character within it." In the year 2000 the proliferation of the most diverse forms of communication—contacts, journeys, transport—would force the world, in Wells's view, to establish a "bilingual compromise" in which each community would use an "ecumenical language," along with its own tongue, which would be limited to its particular sphere.

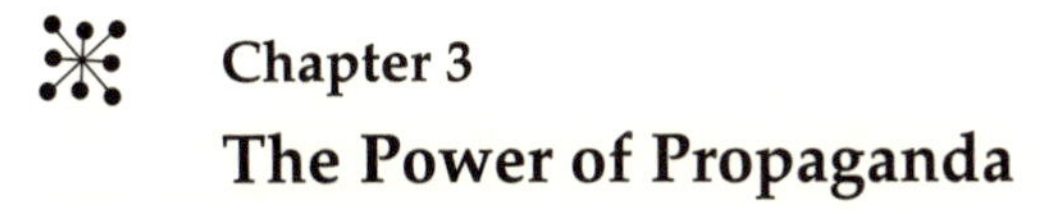

Chapter 3
The Power of Propaganda

World War I conferred a new status on propaganda; peace consecrated it as a method of government. Between the wars the hegemonic ambitions of the United States generated the first fears among European creators about the implications of commercial culture. As the Second World War approached, propaganda strategies set the tone for the internationalization of radio.

Management of Mass Opinion

An Information War

World War I, the first total war, made a priority of what some saw as the mobilization of minds and others as brainwashing. This worldwide political, economic, and ideological conflict spilled beyond the arena of military operations when belligerents created official censorship and propaganda bodies. The most active one abroad was British Crewe House, which employed journalists such as Lord Northcliffe, the owner of the *Times*, and novelists such as H. G. Wells and Rudyard Kipling. As the center for the transmission of dispatches, London became the technological reference for informing the world on the war. In the United States the government set up the Committee on Public Information, otherwise known as the Creel Committee, named after its president, a journalist. It was on this committee that Edward Bernays (1892–1995), the future founder of the public relations industry, first developed his skills. The quantity of rumors, fabricated news, and doctored photographs put into circulation was directly proportional to the severity of censorship.

For France, on a par with other countries in the enforcement of censorship at home, the First World War afforded an opportunity to take stock of the backwardness of its diplomatic apparatus in the field of "intellectual means of action abroad," to quote the expression in vogue at the time. A press club associating journalists and editors was created for correspondents in the various centers of diplomatic activity. In the spring of 1918, a special committee was founded under the aegis of the Minister of Education and Fine Arts, with the mission of guiding "artistic propaganda abroad." Among the more noteworthy members was the Chamber of Haute Couture.

In 1917 in Germany, after its heavy defeat at Verdun, the high command of the Kaiser's army suggested the creation of the Ufa (Universum Film AG). Together with banks and leading corporations, the army grouped dispersed companies and founded a firm whose field of activity encompassed all "sectors of the cinema as well as the manufacture and trade of any activity related to the film industry and projected pictures." The idea was not only to use this enterprise primarily for propaganda purposes, but also to endow the country with a film industry capable of securing a hold over the domestic market dominated by foreign corporations. In February 1916, the government issued several decrees concerning the import of films, henceforth subjected to special authorization. A year later it suspended all imports. For the first time in history, a country countered the free-trade doctrine through the culture industry.

Ufa became the first film company in the world to integrate its activities vertically. The Reich invented the concepts of "cinema for the front," "cinema troops," and "cinema officers." However, the armistice was signed before all the resources of this oversized, overmilitarized project could be mobilized. German strategists were to recognize the effectiveness of the Allies' propaganda as one of the decisive causes of their defeat.

The Revelation of Propaganda

In both belligerent camps, the role of propaganda in the outcome of the war was such that it acquired a reputation of omnipotence. The apologetic discourse of the advertisers and political scientists who founded American sociology of the media transposed this wartime experience into peacetime. The idea developed that democracy could no longer do without modern techniques of "invisible control of society at large," both within the boundaries of the nation-state and beyond. Henceforth, claimed the first specialists in "international relations," diplomacy would have to rely on mass psychology more than on campaigns of gentle persuasion and secret agreements.

In 1922 the American Walter Lippmann (1889–1974) published *Public Opinion*. In this volume, destined to become a textbook in schools of journalism at American universities, he devised a first theory of public opinion in relation to international peace, based on the behavior of the media during and immediately after the war. From his experience as a leader on the propaganda front and an advisor to the U.S. delegation at the peace conference, he drew his first conclusions on the nature of information and the stereotypes preventing governments and their peoples from achieving mutual understanding. He initially tested this theory in "A Test of the News," a lengthy article published as a 42-page report in a supplement to the August 4, 1920 edition of *New Republic*. This work, written in collaboration with his fellow countryman and colleague Charles Merz, also a journalist and former officer, analyzed how, between 1917 and 1920, the *New York Times* constructed the image of the "red peril." Both authors went so far as to talk of a systematic indoctrination campaign aimed at the U.S. public. It was on this type of stereotype that the Department of Justice and the FBI were to base their first witch-hunts against the "agents and conspirators of Moscow," the "Reds," which culminated in the execution of the Italian immigrants Sacco and Vanzetti in 1927. This double execution became a symbol of the miscarriage of justice due to the pressure of galvanized public opinion.

In the same year, Harold Lasswell (1902–1978) published a seminal work that helped to lay the foundations of functionalist sociology of the media: *Propaganda Techniques in the World War.* As the title indicates, his reflection was based on World War I. Propaganda appears in his account as being infallibly effective.

High Culture or Marketing?

As soon as the armistice was signed, the White House dissolved the Creel Committee. It attached no importance to the numerous lessons learned from the war and blocked any attempt to extend the work of official information abroad. The advancing tide of Nazi propaganda thus provoked a particularly rude awakening.

The British government created the Marketing Board, whose mission was to promote products of the empire ("Buy British"). A subsection of the Advertising and Education department was made responsible for film production. The director was John Grierson (1898–1972), a Scot who had fought in the war on a minesweeper and had then gone to the United States to observe the production of Robert Flaherty's first documentary films and the beginnings of the public relations industry. He made his department a breeding ground for a British school of documentary film and called upon foreign filmmakers to take part in it. Grierson was also the strategist behind the creation of the British Council and its network of cultural services, a vast plan of action for the promotion of England, in which movie propaganda was to play a key role.

As for France, it failed to grasp the implications of the rapid growth of audiovisual techniques and the strategic role of information and propaganda. Confident of the universal vocation of the Enlightenment culture, it perpetuated the main lines of its policy of "international cultural relations." Convinced that this influence abroad could be measured in terms of the elite it reached in the targeted countries, French policy concentrated on sending university professors to teach abroad.

The Inexorable Rise of the United States

The Power Base of Communication

The First World War witnessed the development of techniques for coding and decoding secret information and the improvement of telegraph and telephone systems. Above all, however, the war confirmed the role of radio communications and the industrial predominance of Great Britain in this field. During the early postwar years, the U.S. Navy, in the name of strategic national interests, strove to counter this dominant position. At its instigation, American Marconi, a subsidiary of British Marconi, was bought out in 1919 by a consortium of the main players in the electrical equipment and telecommunications industries: General Electric, AT&T, and Westinghouse, subsequently joined by United Fruit Company. This operation led to the creation of a firm specializing in radio communication techniques, RCA (Radio Corporation of America). Starting in 1926, this firm strove to build up the first radio network in the United States (NBC). IT&T (International Telegraph and Telephone), a symbol of the rapid rise of the United States in world networks of distance communication in the late 1930s, was to dislodge British companies from the monopoly they had enjoyed on long-distance calls in South America since the laying of the first undersea cables.

The significance of industrial developments under the aegis of the military was already visible in the international sphere by the end of the 1920s. By that time, the integration of various long-distance transmission techniques was the focus of debate on the regulation of international networks. In 1932, the Telegraph Union and the Radiotelegraph Union merged, creating the International Telecommunications Union. The term "telecommunication," invented by a French engineer at the beginning of the century, appeared officially for the first time. Another first was the use of the term "information" outside the limited context of journalism, with its conversion to a unit of measure in a statistical theory of signals that opened the way to the binary code.

The Specter of Hollywood

By 1919, 90 percent of the films shown in European cinemas were made in the United States. The supremacy of French companies was a thing of the past. After Pathé's withdrawal to its New York subsidiary, its considerable downscaling, and its loss of both foreign and domestic markets, the leading French company would never fully recover from the shock of the war. In 1927 it was transformed into Kodak-Pathé. Against a background of generalized economic recession, the advent of talking pictures compounded the difficulties of the French industry.

U.S. companies were quick to occupy the markets left vacant during the war. The fact that production costs could be recouped on their own domestic market and on the more limited markets of immediate neighbors, proved to be an essential advantage. Considered as a source of additional profits, foreign distribution could be conducted with great price elasticity, especially since the U.S. film industry was restructured around five majors (Paramount, Metro-Goldwyn-Mayer, 20th Century Fox, Warner, and RKO) and had its own modes of consensus when it came to consolidating shares of the export market. The first derived products appeared along with movies, with Walt Disney in the lead. In 1930, scarcely three years after his birth, Mickey Mouse settled into the columns of the *Petit Parisien* with a daily comic strip, and by 1933 had his own *Journal.*

The only film industry that rivaled U.S. producers was Germany's. With the advent of sound, the struggle between the two cinematographic powers turned on patents rather than films. In 1930 the Paris agreement concluded between German and U.S. firms carved up the market into two spheres of influence. Profits from the use of sound film machines were to be reaped only by the financial groups of the two contracting parties. The agreement was based on the one concluded in 1907, between the main players in the world's electrotechnical industry, to limit competition in this highly concentrated sector.

This agreement on equipment hardly made a dent in Hollywood's power. In film production, the majors continued to overwhelm all competition. More and more countries considered taking steps to protect their markets so as to stimulate local production. The Weimar Republic renewed the imperial decision to limit the import of U.S. films. During this short postwar period, with its extraordinary blossoming of the arts in Germany, intellectuals and artists from the world over visited Ufa's Neubabelsberg studios. At the time Germany was also innovating in the field of photojournalism with the creation of a new genre: illustrated news magazines. Yet the financial crisis of 1927 forced the German film industry to compromise with the majors. By signing the Parufamet agreement, U.S. companies agreed to contribute toward German production provided that the quota of American films allowed in the German market be increased. Paradoxically, it was in Hollywood that the most renowned German film producers were able to develop their talents, for the rise to power of the Nazis in 1933 and their total takeover of Ufa two years later forced many producers into exile. Similarly, the authoritarian attitude of the new regime toward the press caused many photographers who had helped to launch the major news magazines to emigrate. The genre created in Germany inspired the French magazine *Vu* and was later adopted by *Life,* founded in 1936.

France also opted for a policy of protectionism for its film industry, the initial stages of what has become a long national tradition. In 1928 the Herriot decree set an annual quota of 120 American films, a figure corresponding to the average national production per annum before the war. In Britain, cinemas were obliged by law to ensure that 30 percent of the full-length films and 25 percent of the short films shown were locally produced. From that time on, the majors stimulated domestic production and sought out more coproductions in an effort to circumvent the quota system, taking advantage of a very loose definition of "national films." English law, for example, considered films produced by Americans, using American directors, scriptwriters, and actors, as national, provided that a set percentage of production costs were paid to British technicians.

Conquest of the international film market by U.S. companies and their advocacy of free trade in this area had their counterpart in the press, where U.S. news agencies took advantage of the weakening of rival positions during the war. This was the case, in particular, of United Press International (UPI), which grasped the opportunity to offer its news services to newspapers in Latin

America, an area that had been allotted to Havas since 1870. Wherever it went, UPI prided itself on pluralistic information, different from that provided in a context of censorship. In 1930, AP and UPI definitively threw off the yoke of the European cartel in the name of a new strategic principle of internationalization that delegitimized the idea of protected territories: free access to information throughout the world.

The First Wave of Advertising

The war transformed the United States from a borrower nation into the world's main creditor. Toward the end of the 1920s, the Fordist economy dislodged British capital from many of its foreign positions, and the dollar replaced the pound as the key currency. A new world economy centered around New York was born. This upward curve of U.S. investments abroad was accompanied by the establishment of advertising agencies, which became the bridgehead of commercial culture.

In 1927 General Motors invited J. Walter Thompson (JWT) to represent it throughout the world and to set up agencies wherever the company built assembly plants and distributed its vehicles. Even as the full force of the recession hit U.S. advertising revenue, JWT founded subsidiary after subsidiary: Antwerp and Madrid in 1927; Paris and Berlin in the following year; Montreal, Bombay, São Paulo, Buenos Aires, Stockholm, and Copenhagen in 1929; Australia and South Africa in 1930, and Toronto and Rio in 1931. After General Motors, Eastman-Kodak, Kellogg's, Ford, RCA, and Chesebrough-Pond's became clients of the agency. A second network, McCann-Erickson, arrived in Paris and London in 1927 and in Berlin in 1928, at the service of a single leading brand: Esso. An ardent supporter of advertising, the writer Blaise Cendrars celebrated it as "an art that calls for internationalism, or polyglotism." Apart from one or two British agencies, there was no agency of another nationality in the international marketplace at the time. France still lagged well behind in an earlier mode of advertising.

It was in New York in the 1920s that Marcel Bleustein-Blanchet, creator of the first modern agency, Publicis, and inventor of radio advertising in France, started his career. At the time, in an

America that had invented the notion of market share and the first systematic market surveys, the marketing industry was in the process of becoming a basic organ of Fordist strategies for managing both a company and social relations in general, through the incorporation of the masses into the nascent consumer society. Following advertising agencies, the first market study and opinion poll agencies were established in the latter half of the 1930s in London and Paris. World War II halted this first wave, at least in those countries involved in the conflict, for elsewhere the expansion of U.S. agencies continued. The main target was Latin America, which combined two advantages: most countries in the region had opted for the commercial organization of their media, and the region as a whole was a prime area for U.S. capital between 1930 and 1950. Only later was the flow of direct investments from the United States redirected toward Europe. It was thus under the auspices of foreign agencies and advertisers that radio serials or *radionovelas,* followed by *telenovelas,* made their debut, well before producers in various Latin American countries appropriated them and gave them an autonomous form.

Yet even before the Second World War was declared, the foundations of a corporate organization with international ambitions were laid. In 1924 an alliance was formed between the association of British advertising agencies and its U.S. counterpart. In 1938 the International Advertising Association (IAA) was founded in New York. It was the first organization for the defense of the professional interests of the three components of this industry: agencies, advertisers, and the media. One of its missions was to make advertising more ethical through adherence to the International Code of Advertising Practices. This code of conduct was developed in 1937 by the International Chamber of Commerce, created in the aftermath of World War I by the directors of the leading European and American firms. Their aim was to participate in the creation of a new world economic order. Conveyed by this professional deontological charter, the ideas of self-regulation and self-discipline, opposed to those of control through measures taken by public authorities, started to penetrate the international marketplace. With them came another key idea that linked democracy to the democratic marketplace, that is, linked citizens' freedom of expression to the "freedom of commercial speech," and freedom itself to the free circulation of commodities. All these arguments were used by the majors, united in the MPEA (Motion Picture Export Association) with the firm intention of abolishing any protectionism that might hinder the free circulation of their films.

Americanization or Crisis of Civilization?

In certain artistic and intellectual circles, the sudden appearance of the United States' financial networks and of products of the entertainment culture on European markets was perceived as an attack on a tradition anchored in high culture. A quarter of a century earlier, H. G. Wells had considered this culture the best guarantee of the perpetuity of French presence abroad. The notions of Americanization and Americanism henceforth stigmatized a foreign threat to the European spirit because they represented a machine-driven industry, gregarious democracy, leveling down, regimentation, and materialism; these were the implications of a confrontation with the "American jungle" and its culture of money. "Americanism is overwhelming us," declared Luigi Pirandello, winner of the Nobel prize for literature in 1934. "I think a new beacon of civilization has been lit over there. The money circulating in the world is American, and it is being followed by life and culture." In a break with this conception of relations between Europe and the New World, marked by the defense of high culture, the philosopher Antonio Gramsci foresaw changes, due to the growing legitimacy of Fordism and its ideal of rationalization in industrial production, that stretched further than the reorganization of corporate life and would affect all the mechanisms of social regulation.

In 1930, in his book *Civilization and Its Discontents*, Sigmund Freud considered the "causes of disillusion" and underscored the ambivalent nature of the "newly won power over space and time." In particular, he questioned the significance of photography and phonograph records as "materializations of the power man possesses of recollection, his memory." "Man," he wrote, "has, as it were, become a prosthetic God. When he puts on all his auxiliary organs he is truly magnificent, but those organs have not grown on to man and they still give him much trouble at times. . . . Future ages will bring with them new and probably unimaginably great advances in this field of civilization and will increase man's likeness to God still more. But in the interests of our investigation, we will not forget that present-day man does not feel happy in his Godlike character."

Inaugurated with the German Oswald Spengler's apocalyptic vision of the "decline of the West," the interwar period closed with Paul Valéry's reflection on the "crisis of the mind." For the French writer, this crisis was part and parcel of that of the European identity and universality, not the product of exogenous factors. It was caused by the loss of "cultural capital" or the growing scarcity of those men who "knew how to read: a lost virtue" and those who "knew how to hear and even to listen," who "knew how to see, read again, hear again, and see again."

Internationalization of the Air Waves

The advent of radio reinforced the internationalization strategies of government propaganda. One country was well ahead of all the others in this respect: the Soviet Union. In 1929 it inaugurated regular programs in German and French and, in the following year, in English and Dutch. This was the logical continuation of the strategy of exporting the revolution, formulated in 1921 at the 3rd Congress of the Communist International, in a programmatic document entitled "Theses on the Organization and Structure of Communist Parties." The founding of Komintern, a centralized world organization, laid the basis for an awesome network of international communication, with the sister parties serving as relays and bases. In 1923 the Soviet party-state reorganized its news agency, henceforth known as Tass.

In 1931 the Roman Catholic Church also endowed itself with a multilingual communication device: Radio Vatican. But it was Germany that set in motion the real dynamic of radiophonic internationalization. The Nazi government forged the concept of psychological warfare, a leitmotiv of *Mein Kampf,* and applied it to its foreign policy. In 1933 the Zeesen shortwave station, near Berlin, was inaugurated. Its broadcasts were aimed at numerous immigrant German communities and, in English, at the United States. Three years later, during the Olympic Games in Berlin, it was capable of broadcasting in twenty-eight languages. In 1935 Mussolinian fascism proved to be equally precocious in its use of radio for political proselytism, by broadcasting in Arabic to Africa and the Middle East. In 1936, during the Spanish civil war, the use of foreign-language radio broadcasts by both camps highlighted the strategic role of this new form of propaganda.

The first retaliation to broadcasts from Berlin came in 1934 when the Social-Christian chancellor of Austria, Engelbert Dollfuss, ordered the jamming of the radiophonic spectrum shortly before his assassination by the Nazis. Repeated efforts by the League of Nations to rally the various countries in the international community to signing pacts of "radiophonic nonaggression" (the first convention was signed in 1936 by most of the member states) were doomed to failure. In this domain as in many others, the League of Nations was unable to impose itself as the "court of public opinion" or the "conscience of the world," to quote one of its initiators, President Woodrow Wilson. (Paradoxically, the United States, converted to isolationism, obstinately refused to belong to it.) The International Broadcasting Union, founded in 1925, was hardly more effective. This organization, under the predominant influence of Germany, was the only international institution not to suspend its activities during the Second World War. Frequent use of the radio for propaganda purposes did not preclude the ideologies of communication as redemption. Lewis Mumford (1895–1990), a U.S. historian of technology and urbanization, while quite lucid about the potential of this technology for "mass regimentation," perpetuated the utopias of Kropotkin and Geddes by conceiving of another possible use of broadcasting networks: as a means of reviving the agora of the small ancient Greek city-states.

It was only much later that Great Britain and the United States became aware of the strategic importance of the national-socialist state's propaganda networks. In 1938, the BBC—which was to be instrumental in the struggle against the Axis countries by counterbalancing their power with its own broadcasts in twenty-three different languages—created first a German-language service and then began broadcasting in Spanish and Portuguese to Latin America. In the same year the White House undertook to mobilize private broadcasting networks in the United States for the purpose of neutralizing the growing influence of Germany in Latin American countries. It was aiming primarily at the large colonies of German immigrants who were particularly active in spreading the ideals of Hitler's regime. Walt Disney productions, *Time* and *Life* magazines, and the *Reader's Digest* all joined the antifascist cause. Founded in 1922, the *Reader's Digest* was one of the first periodicals to publish foreign language editions, beginning in 1940. These were initially in Spanish and Portuguese, specifically intended to counter the influence of the Axis in Latin America. It was also for South American countries that *Time* published its first regional editions in English.

In February 1942, Washington took over from private broadcasting companies and created an official radio station, the *Voice of America.* Propaganda abroad was placed under the responsibility of two bodies: the Office of War Information (OWI), for overt propaganda, and the Office of Strategic Service (OSS), in charge of covert propaganda. Unlike the previous war, when it was mainly journalists and writers who committed themselves to the cause, these new propaganda institutions recruited their specialists from among professionals in advertising and public relations agencies, and sociologists, psychologists, and anthropologists from academic milieus. That was how most of the pioneers of functionalist sociology of the media became familiar with international realities.

Chapter 4
The Bipolar Geopolitics of Technology

The Cold War perpetuated the concept of communication as propaganda, and its strategic scenarios guided the establishment of satellite systems. As the harnessing of the third world became a major issue in the confrontation between the two political systems, north/south issues assumed an increasing role in the east/west rivalry. In the struggle against underdevelopment, communication became synonymous with modernization.

Winning Hearts and Minds

After the Second World War, the OSS (Office of Strategic Services) was transformed into the CIA (Central Intelligence Agency), and the OWI (Office of War Information) became the Office of International Information and then the USIA (U.S. Information Agency) in 1953. Washington added two radio stations: Radio Free Europe in 1950, which broadcast to countries in the Eastern bloc, and Radio Liberty in 1953, which was aimed at audiences in the Soviet Union. From the outset both were funded by the CIA and, unlike Voice of America, whose journalists were American, were operated by immigrants from the socialist bloc. Despite Stalin's willingness to dissolve Komintern in 1944 in exchange for support from the United States and England, and his replacement of the hymn of the worker's International by a hagiographic national anthem to his own glory, Radio Moscow continued to broadcast its party-state propaganda abroad.

There were two radically opposed theories on the regulation of international flows of information. On the one hand, the principle of the "free flow of information," derived from the liberal doctrine of the free circulation of goods, was ratified by Congress and elevated to the rank of official doctrine by the U.S. Department of State, which included it in its war objectives starting in 1944. On the other hand, the Kremlin doctrine aimed to prevent Soviet citizens from having any contact with Western media, concealed its internal security needs behind the notion of national sovereignty, and interpreted unwanted international broadcasts as "interference by a foreign power in the domestic affairs of a nation-state." The idea of "ideological aggression," as old as the revolution, complemented the Soviet Union's self-representation as a besieged citadel.

In 1947 the International Telecommunications Union (ITU) was attached as a technical body—like the Universal Postal Union—to the new system of the United Nations. Debate on the uses of the airwaves took place under the ITU's aegis. The Atlantic City conference, held in the same year, simply ratified the "prior claim to use" concerning the distribution of frequencies, which had been adopted by the world's dominant naval powers at the beginning of the century. Three years later, the Soviet viewpoint, which legitimized the systematic jamming by states of foreign broadcasts, was defeated. This was merely the start of a debate that was to last until the Berlin wall fell. Blowing hot and cold, Moscow and its satellites sought to penalize anyone who listened to foreign programs and to block the development of the civil radio communication industry.

On both sides of the Iron Curtain, theories of plots and manipulation became catechismal references for deciphering the opponent's deeds and gestures. Each party saw itself as being engaged in a battle to "conquer hearts and minds," to quote the phrase coined by the experts in psychological warfare. In early 1953, functionalist sociology of the media in the United States, recognizing the importance of this question, founded a new field of study officially known as international communication. The Manichaean character of the hypotheses formulated by many of its practitioners was largely due to their background, for many of them had worked for the OSS and the OWI as experts in psychological warfare during the Second World War. In fact some of them resumed this type of work as permanent advisors to the Voice of America or to U.S. troops during the Korean War (1950–1953). That was the case, notably, of the future founder of the famous Institute for Communication Research at Stanford University, Wilbur Schramm. In 1951 he coauthored the book *The Reds Take the City*, on the role of psychological operations in the occupation of Seoul by communist armies.

In the east, doctrinarians of the party-state categorized the agents of ideological aggression in absolute terms by labeling all Western media "bourgeois means of propaganda." The linguistic plurality of Soviet broadcasts abroad (e.g., in 1970, 235 hours per week of broadcasting in fifteen languages to Africa, as opposed to 130 hours in four languages by the Voice of America) did not, however, offset the univocal character of their wooden language. Increasingly out of step with the realities experienced by their audiences, even in the view of USIA advisors, the Soviet media were already meaningless—except for those who had always been convinced.

The Conquest of Space

The Military-Industrial Complex

For the Soviet Union the imperative of defense against the foreign threat was, from the regime's inception, one of its main legitimizing arguments and one of the main driving forces of the economy, but for the United States it represented a novelty caused by its entry into the Cold War.

From 1947 this unusual situation was institutionalized by the National Security Act, which provided the legal framework for maintaining the exceptional mobilization of the war years and prevented a return to the apathy that had characterized the depression in the 1930s. It broke down divisions between private and public, civil and military, applied research and basic research, industrial laboratories and universities, thereby renewing the experiences in synergy that had proven their value during the Second World War and had spawned, for example, the development of ballistic systems based on large computers of the first generation. Here was a sign, among others, of the massive participation of the U.S. state in research and development expenses of electronic and aerospace firms, the cradle of information and communication technologies. In 1930 the federal budget's share in private and public research was a mere 14 percent; by 1947 it had shot up to 56 percent. The inventor of cybernetics, Norbert Wiener, in 1948 saw the advent of the "information society" as a guarantee that there would be no return to the barbarity of World War II; the new logic of global confrontation would soon make it necessary to tone down this analysis.

The flow of funds from the Pentagon, renewed with each stage in the Asian wars, played a decisive part in the invention of the first transistor-based computers by IBM in 1959. Above all, these subsidies made possible the construction of the first intercontinental networks, initially conceived in a context of close coordination between military needs and industrial production. During the same decade, the continental defense network SAGE (Semi-Automatic Ground Environment) was built for the U.S. Air Force. By connecting each computer to a radar registering flight trajectories and by establishing a telephone link between all the computers of the system, this apparatus brought into existence the notion of real-time data transmission, i.e., teleinformatics or data communications. In 1958 the first experimental computer links were established between laboratories working for the U.S. Department of Defense. By 1968 the first data transmission network, ARPANET (Advanced Research Project Agency Network), had been set up under the same auspices. It linked up the computer centers of universities that, in turn, were linked to London by satellite and to areas of the Pacific via Hawaii. From its inception in the context of state security, this system maintained the original idea of a network of computers linked up in such a way that digital data could be transmitted via several different routes. The system as a whole would therefore not be impaired by the destruction of one or more computing centers. In the 1990s it was to serve as an essential reference for the Internet.

Intelsat

The first artificial satellite, *Sputnik,* launched in 1957 by the Soviet Union, opened a new front in the Cold War: the space race. To meet this challenge, President Eisenhower founded NASA (National Aeronautics and Space Administration) in the following year. The main aim was not only to conquer the moon, but also to establish a worldwide communication network.

In 1962 the satellite *Telstar* linked Europe to the United States. In 1965 *Early Bird* was put into orbit. It was the first geostationary commercial telecommunications satellite of the International Telecommunications Satellite Consortium, otherwise known as Intelsat. This consortium was the institutional form, approved by Congress, that NASA and the U.S. aerospace industry proposed to countries of the "free world," from 1964, as a means of creating a system of global communication. Initially the United States had complete ascendancy over Intelsat. It was managed by a peculiar kind of private company, Comsat; 45 percent of the shares were owned by AT&T, ITT, RCA, and GTE. The other half was spread out among a myriad of small shareholders and the remaining 163 firms in the U.S. communication industry. Three members of the board of directors were White House delegates. Furthermore, the United States owned 60 percent of Intelsat, a quota set according to the actual use made of it by each country. The United Kingdom, France, and West Germany owned 20 percent and the rest was shared between fifteen other industrial nations. No third world country was a partner in the consortium. Yet, faithful to the doctrine of the international welfare state, in which all should have equal access to the system of world communication, Intelsat offered all developing countries a reduced rate subsidized by the industrialized nations.

In 1965 the Soviet Union proposed access to its own system, Intercosmos, to its partners in the socialist camp. Six years later it set up a commercial organization, Intersputnik.

The space race, as a grand narrative of a New Frontier, lasted a little more than ten years. In the era of détente, projects for coupling up the vessels of the two superpowers (Soyuz) flourished. The U.S. space industry initiated its redeployment by favoring applications with short-term goals. Although the Pentagon's spending on spy satellites remained stable, the bulk of NASA's budget was henceforth to be devoted to the launching of satellites for communication, weather observation, air and maritime navigation assistance, and observation of natural resources. In July 1972, NASA launched the first civil earth observation satellite, ERTS-1 (Earth Resources Technology Satellite), subsequently renamed Landsat-1.

The Soviet authorities, by contrast, did not think in terms of deriving products for civilian use from their military systems. Based on the withholding of information, the political system continued to be spurred by the priority and exclusive logic of defense. Although the Soviet industrial machine was capable of producing, in 1947, the famous Kalachnikov, the weapon used by guerrilla fighters throughout the world, it was incapable of conceiving an object such as the transistor, which, once popularized in the latter half of the fifties, was to change daily life and geopolitics. To prevent its citizens from listening to foreign broadcasts, the Soviet authorities promoted collective listening and produced wired receivers that were unable to pick up foreign radio stations.

For a long time, the conquest of space was controlled solely by the United States and the Soviet Union. It was only in the 1980s that Europe managed to compete with the U.S. communication satellite and launching industries. The United States' reaction was immediate. President Reagan deregulated the intergovernmental Intelsat, placing it in direct competition with private satellites and deleting the clause granting reduced rates to third world countries. In other areas of space technology applications, successful launches of the first civil earth observation satellites (Spot-1, 2, and 3) took place between 1986 and 1993. In the military field, European dependence on U.S. Secret Service satellites, such as *Keyhold* and *Lacrosse,* lasted throughout the 1970s and 1980s. It was particularly clear during the Gulf War, in 1990–1991, and in Bosnia, but Europe began to overcome this dependence in 1995, when the first spy satellite of the Helios program was launched as the initial stage of a European space network project for strategic intelligence gathering. In the meantime, the club of non-

European space powers had also grown, incorporating, in particular, China and India.

Integrating the Third World

Communication for Development

In 1949, in a speech on the state of the Union known as "Point Four," President Harry Truman prioritized the struggle against underdevelopment. The notion of development, which before the war referred primarily to the degree of "culture" and "civilization" achieved by a nation, took on an economic connotation. It served to *direct* a vast program for mobilizing energy and public opinion in a struggle against the great imbalances seen as likely to favor "world communism." Inaugurated in the 1950s in the ultrasensitive areas of the Middle East where attempts were being made to nationalize oil wells, this program was fully deployed in the following decade in Latin America. To counter the Cuban revolution (1959), Washington offered its southern neighbors the Alliance for Progress, presented as a revolution with freedom.

Sociologists at American universities derived numerous hypotheses from the lessons learned from their participation in wartime psychological operations abroad and from the progress made by industrial marketing at home. The question of development was defined as a process of diffusion of innovation. The goal of strategies of persuasion was to stimulate change in the attitudes of "underdeveloped" populations; that is to say, to lead them from a so-called traditional culture and society to a so-called modern one. The Westernization ideal represented all the qualities characteristic of a "modern attitude" and "cosmopolitan tastes."

Indexes of modernization were calculated in terms of literacy, industrialization, urbanization, and exposure to the media. The curves and typologies thus established placed third world countries on a scale of growth in revenue per capita. For almost a quarter of a century, this frame of reference governed all understanding of north/south relations; it fit the spirit of the times. It lay at the heart of governmental development aid policies and impregnated the philosophy of the United Nations. UNESCO hastened to translate the basic texts of this instrumental sociology into several languages, while its staff established catalogs of minimal standards: to extricate itself from underdevelopment, to "take off," a country had to have ten copies of newspapers, five wirelesses, two television sets and two cinema seats for every one hundred inhabitants. As vehicles of modern behavior, the media were seen as key agents of innovation. As messengers of the "revolution of rising expectations," they propagated the models of consumption and aspirations symbolized by those societies that had already attained the higher stage of evolution. This absolute belief in exponential progress and in the modernizing virtue of the media merely updated ethnocentric conceptions of nineteenth-century diffusionist theories. "Primitive" societies were now termed "underdeveloped," and their only option was to imitate the models of their elders. This conception permeated the use of the audiovisual media for the purpose of "rationalizing" the attitudes of peasants (e.g., toward the adoption of technology or the use of fertilizers) and the behavior of working-class women regarding birth control. Relayed by local authorities, it stimulated experiments in the use of satellites for educational purposes in very large countries such as India and Brazil. An extreme case in point was the Brazilian military dictatorship, which refused to renew the literacy campaigns based on mass mobilization initiated by the regime it had overthrown and called upon education engineers from Stanford, in the 1970s, to launch short-lived satellite experiments in the poorest area of the Nordeste. At the same time, it rejected its own sociologists, educators, and anthropologists, forcing them underground or into exile.

Revolt

In April 1955, the Bandung Afro-Asian Conference (Indonesia) initiated the Non-Aligned Movement. Three years earlier, demographer Alfred Sauvy and anthropologist Georges Balandier had given the name "third world" to this international "third estate." Radio became a weapon in the hands of liberation movements. The most well known was the voice of the pan-Arab revolution (The Voice of the Arabs), inaugurated in 1953 in Egypt by Colonel Nasser's regime. Broadcasting from Cairo, it became the mouthpiece of the pan-Arab revolution. In 1956, La Voix de l'Algérie combattante (The Voice of Fighting Algeria) influenced Algerian society with broadcasts from Tunis. The French authorities retaliated by systematically jamming its programs and prohibiting the sale of radios and even batteries. Two years later, Radio Rebelde (Rebel Radio) of the Castroist guerrillas, at the initiative of Ernesto Che Guevara himself, launched its first message from the freed territories of the Sierra Maestra.

The Algerian conflict proved to be a lesson in media practices. French counterinsurrection specialists discovered the role that the media had come to play in legitimizing the objectives of a revolutionary movement vis-à-vis public opinion. Appalled by the behavior of certain media in metropolitan France and of the international press during the Algerian war, they accused them of having played the enemy's game.

A few years later, at the end of another counterrevolutionary war, in Vietnam, many advisors of the Pentagon argued in similar terms. Despite substantial investments in overt and covert psychological operations, and other "pacification" campaigns, propaganda strategies were unable to outweigh public opinion. That, at least, was what political scientist Samuel P. Huntington concluded in 1975, the year Saigon fell. In particular, he blamed the new source of national power composed of networks, news magazines, the *Washington Post,* and the *New York Times.* Extrapolating lessons from this war, Huntington incriminated the media's freedom of tone and accused them of being a chief cause of the crisis of ungovernability in Western democracies. These very notions were featured in the title of the report he wrote with Michel Crozier of France and Joji Watanuki of Japan, for the Trilateral Commission. Formulating solutions to this crisis was the main purpose of this group of supposed private citizens, founded in July 1973 at the initiative of David Rockefeller, president of Chase Manhattan Bank, and composed of over two hundred personalities from the three regions of the Triad (North America, Western Europe, and Japan).

Chapter 5
Transnationalization and Geoeconomic Rationality

The geopolitical representation of the world, nurtured by the Cold War, had the effect of weakening geoeconomic logics. The force with which these logics structured the international sphere did not become obvious until the 1970s. By questioning the international economic and communication order, the new historical subject constituted by the third world revealed the numerous manifestations of unequal exchange. European nation-states began to formulate a reaction to the threat of destabilization of their cultural and technological policies by multinational corporations.

Toward the End of the Monopoly of Interstate Relations

The networks of nonstate interactions and transactions, as agents of worldwide integration, gained increasing importance as a result of two types of processes: first, the formulation of industrial policies in the context of regional integration, primarily in the European Community; secondly, the opposition of major foreign companies to third world countries' attempts to nationalize the strategic sectors of their economies. An extreme case served to reveal this trend: the strategy of economic and ideological encirclement of the Chilean president Salvador Allende's socialist regime, between 1970 and 1973, by an objective alliance between the forces of internal opposition, the armed forces, multinational corporations (including IT&T), and U.S. intelligence agencies. In the months following the military coup d'état, this collusion became public knowledge through the confessions of its protagonists at hearings organized before ad hoc commissions of the U.S. Senate.

The schemas for analyzing the trend toward worldwide integration came, progressively, to include nonstate, transnational, and transgovernmental actors and forms of interaction. Communication networks were in the foreground of this reformulation; only the new actors needed to be defined. Because the foreign activities of most manufacturing firms were linked to the development of their exports, the notions of international enterprise and internationalization had for a long time seemed to suffice for describing the expansion of these firms outside their home countries. As early as the 1960s, however, these notions proved inadequate for understanding the actors who fed the flow of investments by setting up operations abroad—hence, the birth of the notion of multinational corporations.

The new awareness of the part played by these corporations in international policy and economics changed the way of defining the issue. Asked to study ways of regulating the activities of foreign firms, United Nations experts suggested replacing the term "multinational" with "transnational." "Multinational" gives the impression that these firms were both the sum of several nationalities and each nationality in particular; in short, that they were profoundly rooted in the host country. By opting for the term "transnational," the experts wanted to mark a political distinction, stressing the territorial noncoincidence and the centralized mode of management of these companies. Striving for flexibility, a source of good performance, the transnational firm takes advantage of favorable conditions—natural, financial, political, and legal—prevailing in each host country and tries to avoid those conditions seen as unfavorable to its interests. The term "multinational" eliminated the controversial nature of the expansion of these new units of supranational capital by making the world economy a mosaic of local economies. The term "transnational," which implies the existence of a broad movement toward integration at the international level, is taken to mean that there is a potential source of conflict between the interests of macro-enterprises and the territories in which they operate. In 1974, aware of these implications, the United Nations created a commission on transnational companies, attached to the Economic and Social Council, and a center for research on the same subject, which reported directly to the Secretariat. Their mission was clear. Formulated in administrative terms, they had to "facilitate the conclusion of effective international arrangements concerning the activities of transnational companies, with a view to favoring their contribution both to the national aim of development and to worldwide economic growth, while simultaneously controlling and eliminating their negative effects." The center was thus called on to produce expertise on the strategies of a wide range of enterprises, from pharmaceutical and agro-industrial food companies to advertising networks and transborder data flows.

Pragmatic and well removed from the conceptual controversy, international marketing manuals classified firms with foreign operations in three main categories, corresponding to their approach to market penetration. A firm is said to be "ethnocentric" (or "monocentric") when its foreign subsidiaries are closely bound to the national identity of the mother company. A "geocentric" firm is one whose subsidiaries are "strongly integrated into the search for an optimum strategy in a cosmopolitan perspective." Finally, a "polycentric" firm has few foreign subsidiaries; they are well integrated but are managed on a decentralized basis. This nomenclature obviously covers multiple forms of transnationalization that have evolved over time and applies to different sectors of economic activity.

Transnational companies in the communications sector were among the first to take note of the conflictual relationship between the local, national, and transnational levels. Active in the ultrasensitive area of particular identities, they tried either to avoid them or to adapt to them, like the good followers of social Darwinism that they soon learned to be.

Balance of Power and National Mediation

The Expansion of Advertising Networks

While the Marshall Plan helped Europe, ravaged by the war, back onto the path of growth, it was also the Trojan horse that bore the "Americanization of society." It opened the way to the modernization of the industrial apparatus of the countries concerned and served as a background to the reorganization of society. As Luc Boltanski explains in his work on managers, it was through this breach that systems of values, social technologies, and models of excellence, tried and tested in the United States, flowed in. Human engineering and management became important aspects of business training.

The modernization of advertising in postwar Europe was a feature of these structural changes. Its successive phases illustrate the twists and turns of the process of transnationalization of marketing. Advertising, which initially seemed little more than a modernized sales technique, gradually became the vector of the commercialization of the entire mode of communication and, as such, a key feature of the public sphere. As the most significant locus of the production of "technical events," that is to say, events created from visual or sound artifacts, and which abruptly sever the continuity of a piece of information and restimulate the audience's attention, it was the avant-garde laboratory of mass culture.

With the exception of the networks founded during the Great Depression, the first major wave of internationalization of advertising agencies in Europe started in the 1950s and reached its peak in the following decade, the era of the "American challenge," to quote the title of a best-seller written in 1967 by Frenchman Jean-Jacques Servan-Schreiber. This was an imperial phase in which the main actors were from U.S. firms making direct investments in industry. In most countries where their subsidiaries were established, their arrival in force decimated local agencies. Faced with this reinforced U.S. presence, France was the only European country to maintain a majority share of its domestic market owing to its two historical figureheads, Havas and Publicis. At the time, subsidiaries of U.S. agencies worked mainly for clients of their own nationality; in many countries, and particularly in France, they received no business from the largest state-owned firms and public institutions. Very little interaction existed between the various national subsidiaries. Throughout the world, these companies constituted centers for learning skills that they alone possessed. Encouraged by this omnipresence, *Advertising Age,* the mouthpiece of U.S. agencies, awarded them the title of unofficial diplomats of the nation, for they represented the lifestyle of the country far more intensely and realistically than the State Department or embassies.

The second generation of international networks was born in the 1970s. During this decade of consolidation of national advertising markets, those local agencies that had survived hampered U.S. networks by vying for the same clients and timidly beginning their own internationalization. The growth of domestic markets created a new balance of forces between local professionals and U.S. agencies. Moreover, throughout the world governments were adopting measures against foreign agencies to protect their job markets and safeguard their national languages and cultures and even national moral standards. Faced with what they interpreted as new forms of nationalism, U.S. agencies proposed partnerships with minority participation and recruited local talent. They furthermore began to take into account the existence of "cultural differences." Coordination between national subsidiaries in managing the budgets of transnational firms at the regional or international level was, however, still the exception confirming the rule of juxtaposed agencies. It was only with the advent of global networks in the 1980s that an overall plan emerged. This third generation truly warrants being called the age of networks and geostrategic actors.

A Strategy of Institutional Resistance: French Cinema

In return for the economic aid provided by the Marshall Plan, U.S. negotiators asked the French government to reduce the restrictions on the import of American films that had been applied since the late 1920s. In May 1946 the Blum-Byrnes agreement—named after the French representative, Léon Blum, and the U.S.

Secretary of State, James Byrnes—was signed in Washington. The agreement annulled the measures of the Herriot decree. It replaced the import quota by a so-called screen quota that reserved cinema screens for French films four weeks in every quarter. This was indeed a substantial regression with respect to the previous measure because it allowed French films on the screens 31 percent of the time compared to up to 50 percent before the war. The new quota did not allow for all the potential French production to be shown in cinemas. In 1946 France managed to produce ninety-six films; in the following year this figure dropped to seventy-four. The crisis affected the entire sector, throwing actors, directors, and producers into the streets to resist the threat of losing their livelihood. The support they received from the press forced the National Assembly to reconsider the agreements. New negotiations with Washington resulted in an amended agreement signed in September 1948. It restored the system of import quotas while maintaining screen time criteria. Of the 186 films allowed entry into the country per year, 121 could come from the United States. Screen time for French productions rose from four to five weeks, an increase from 31 to 38 percent. The import of films other than those of U.S. origin was subjected to severe restrictions. Only sixty-five per year were allowed, causing an outcry among British producers. These protectionist measures were accompanied in 1948 by a resolute strategy to encourage film production. At the heart of the new system of support was the *Centre national de la cinématographie* (CNC). One of its main missions was to ensure that a portion of the earnings obtained in France by foreign films was reinvested in national production.

As a result of this strategy of protection and production of national films, France became one of the rare countries in Europe and in the world to maintain a degree of pluralism on its screens. By making the opposite choice, Britain saw its film production virtually destroyed. Its film industry was saved only because its major clients had for a long time been the producers of advertising films. The only solution for British producers such as Adrian Lyne, Tony Scott, Allan Parker, and Ridley Scott was to emigrate to California to make their films, after gaining experience in producing ads in their own country. Italy, which had managed to keep its national film industry alive through an official policy of support for local production, was powerless to stop its decline during the 1980s due to the deregulation and privatization of the audiovisual sector.

The Flexibility of Magazines

The war had propelled *Time* and *Newsweek* into international orbit. In 1946 the former could boast of fourteen editions and the latter of five, though all in English. Only *Reader's Digest* had opted for publishing editions in different languages. Latin American editions were followed by editions for France, Spain, and Portugal, and Sweden and Finland. For a long time this magazine remained the paragon of the borderless editorial product. Its editors soon learned, moreover, to adapt the content to suit the diversity of national contexts, with a skillful mix of articles (composed by a world coordination center located in a suburb of New York), adaptations to local interests and culture, and material produced regionally or locally. Forty years after the first foreign edition, over thirty million copies circulated in about forty editions published in approximately twenty languages. For nearly a quarter century, *Time, Newsweek,* and *Reader's Digest* were the only publications to defy national borders.

Other magazines, including *Scientific American, Cosmopolitan, Family Circle, Playboy, Glamour,* and *Good Housekeeping,* internationalized their publication in the 1960s by means of the so-called franchise system. The company that owned the title sold the right to its use to a national publisher under specific conditions, including the payment of royalties. This system, an outcome of strategic reflection on the tension between national and transnational interests, allowed the franchisee to be linked up to a network, sharing common know-how, a bank of articles and a list of advertisers, and sometimes to participate in brainstorming sessions with the editorial teams of other local editions. The process was a flexible one in which each publication established specific forms of partnership with the head office, which nonetheless maintained strict control over the publication's evolution. A single publication did not expand into all countries at once; there were outposts and rearguards. Certain third world countries had editions even before some major industrial countries (eight years separated the first Latin American editions of *Cosmopolitan*, launched in 1966, and its French equivalent); in other countries there were no editions at all. These local editions were aimed essentially at the middle classes and sought to satisfy more exclusive tastes.

In the 1970s the diversification of national bases of editorial production caused numerous publications from the United States to face competition with similar genres created locally. By the end of the decade French and German groups started to carve out a place for themselves in the world market. *Elle* created several foreign editions and achieved the impossible: the conquest of the U.S. market itself. The core target everywhere was a "young, Western, and urban" audience. The financial press also began its transnational expansion. In 1976 the *Wall Street Journal* launched an Asian edition in Hong Kong. The first European edition of the *Financial Times* was produced in 1979. Its American rival followed suit four years later.

The Awakening of Planetary Consciousness

Toward a New World Order of Information and Communication

The 1970s marked a historical watershed in critical approaches to the industrial mechanisms governing the production of information and mass culture as well as the international imbalances in trade and communication flows. In fact, criticism flourished.

One of its first sources was the Non-Aligned Movement. Its fourth summit, held in Algiers in 1973, laid the foundations for the demand for a new world order of information and communication. The main platform for debate was UNESCO, whose mission is to represent the community of nations in the field of culture, communication, education, and science. These debates were symmetrical to those conducted by the Group of 77 (the equivalent of the nonaligned movement in the economic domain) in the UN General Assembly since 1974. The purpose of the latter debates was to obtain a revision of the international system of trade and lay the foundations of a "new economic order" (including reform of monetary institutions, a significant transfer of resources to finance development, access to markets in the north, and technology transfer). In the area of communication, a key idea underlay this approach: cultural imperialism was a reality, and the situation of cultural dependency that it spawned was not a matter of manipulation or conspiracy, but a structural fact. The effects of domination were at the roots of the principle of unequal exchange between the center and the periphery.

In 1969 UNESCO, then presided over by Frenchman Jean Maheu, called a meeting of experts in Montreal at the request of its member-states. The intention was to draw up an inventory of knowledge on the subject and define new paths for research. At the center of this meeting was a debate on the one-way communication characterizing relations between developing countries and the others, which was likely to lead to problems of understanding between nations. In 1972, at the initiative of the Soviet delegation, a proposal for a convention to regulate the transmission of direct broadcast satellites (which do not require ground relays) was submitted first to UNESCO and then to the UN General Assembly. In the vote on this principle, the United States found itself completely isolated.

The debate on the new order began with criticism of the often tendentious, incorrect, nonobjective, and unsuitable coverage by the four main news agencies in the developed countries that monopolized the international dissemination of news. U.S. agencies, which had demonstrated their partiality through their opposition to the regime of the Chilean president, Salvador Allende, were among the chief targets. The debate became a forum for grievances on issues ranging from the allocation of radio frequencies to the construction of national communication infrastructures. In 1977 a report was commissioned by the new director of UNESCO, Amadou Mahtar M'Bow of Senegal. The task was entrusted to an international commission for research on communication problems, presided over by Irishman Sean MacBride, a winner of the Nobel and Lenin peace prizes. The final report, published in 1980, was the first official document published by an international body that clearly raised the question of the unbalanced flows of agency dispatches, television programs, films, and other cultural products, as well as equipment.

The MacBride report was the subject of extensive debate at the UNESCO general conference held in Belgrade in late 1980.

Numerous factors led to a stalemate. Intransigent in the extreme, the Reagan administration strove at all costs to impose its intangible doctrine of the "free flow of information," while the communist bloc countries confused the issue by supporting the legitimate demand for the cultural emancipation of the south while steadfastly refusing to open their own mass communication systems. For authorities in the Eastern bloc, the moment was particularly crucial. To the specter of the direct broadcast satellite was added the palpable reality of a system of social control that, despite all the measures taken to jam the airwaves, was made increasingly porous by cross-border media. Relayed locally by dissident movements, Western radio stations and TV channels, and a little later videos, insidiously undermined the system on a daily basis by reflecting a lifestyle contradicting the economy of dearth and the slogans chanted by party-state propaganda. Finally, the nonaligned countries were characterized not only by extreme technological heterogeneity but also by internal political contradictions. Some regimes used this international tribune to point to foreign scapegoats and to try to clear their own names with regard to their shortcomings in granting freedom of expression to journalists and creators. Notwithstanding these numerous limits, the debates were the first cry of alarm concerning the unequal exchange of images and information. Apart from their sometimes virulent rhetoric, they were the expression of protest against the prevailing model of development, and hence against a type of relationship between north and south, materialized since the 1950s in modernizing strategies. Critical of the vertical schemata of communication dictated by diffusionist theory, they raised the question of the relationship between democracy and development and between communication and citizens' direct participation in their own development. They also afforded an opportunity to bring up the buried memory of philosophies and thinkers from the third world who opposed the productivist and rationalist view of development. In terms of specific measures, the philosophy of the new order inspired the creation of national news agencies or pools of regional agencies. Its recommendations also gave rise to sectoral policies such as those mentioned above, aimed at regulating foreign advertising agencies or establishing reserved markets and quota systems to support national film production.

In 1985 the United States, denouncing the drift toward the "politicizing" of communication issues, walked out of UNESCO. It was soon followed by Singapore and England, then governed by Margaret Thatcher. Washington threatened, for the same reasons, to leave the International Telecommunications Union, which from 1979 had been responsible for organizing the international administrative conference on radio. For the first time, and to the great displeasure of the United States, 142 delegations were urged to change the rule allowing countries to claim frequencies on a first-come basis—a rule imposed at the beginning of the century by the handful of major maritime powers.

Discussions on the new order undoubtedly had the function of awakening strategic consciousness. For corporate lobbies, such as the International Advertising Association (IAA) and the Inter-American Press Association (IAPA—a press owners' association), which were among the most active in defending corporate interests, it was the starting point of an institutional reorganization with a view to meeting the challenge that they identified as global and far from conjunctural. A similar awakening took place among nongovernmental organizations that, seeking to go beyond mere verbal commitments by states, took initiatives on the fringes of official institutions. A case in point was the networks of NGOs that, between 1974 and 1976, successfully orchestrated the international boycott of Nestlé in order to curb its obtrusive advertising and promotion campaigns for breast milk substitutes in the third world.

Europe: The Reverse Side of Cultural Policies

The second source of a doctrine on the consequences of the internationalization of cultural products was Europe, where France played a key role.

In late 1978, for the first time, the European ministers responsible for cultural affairs spoke explicitly of culture industries that were multinational by nature and recognized that the legal regulations set up by the nation-state to curb them carried very little weight. Shortly before that, the notion of culture industries had been introduced, by experts from the French Ministry of Culture, into the terms used by the Council of Europe. This notion was based on the premise of an unequal battle between public policies aiming for the democratization of cultural goods and the inexorable rise of another form of democratization by the market, that is, through the products of mass culture. It was also an observation that a threat was posed to the national identity as a result of growing breaches in the borders of the nation-state.

In the 1970s no links were established between the cry of alarm from the south and the warnings of those in charge of culture in Europe. It was only with the economic upturn during the early years of the socialist presidency in France that a European government called openly for a policy capable of "guaranteeing the countries of the south the capacity to control their means of communication and the messages they convey" (speech by President François Mitterrand to the summit of the seven most industrialized countries, in June 1982). It also proposed a "true crusade against the domination of financial and intellectual imperialism" (speech by the French minister of culture, Jack Lang, to the UNESCO world conference on cultural policies, held in Mexico in July 1982). In October 1981, during the North/South summit in Cancún (Mexico), the French president had asserted that "the free market allows for no growth other than that of the multinational firms, which create streams of wealth and oceans of poverty in the third world."

By the end of the 1970s, however, one thing was already clear: the answers to new problems provided by international institutions were likely to fall far short of the expectations of all parties concerned. Few recommendations ever reached concrete form. A case in point was the code of conduct for transnational corporations, drawn up by the United Nations. The increasing legitimacy of neoliberal policies, which were averse to any kind of control or regulation of the activities of transnational firms by public bodies, definitively sealed this code's fate in the 1980s. The commission and the center responsible for formulating a regulatory framework were, moreover, dissolved. When discussions did, in some cases, result in the adoption of a code by governmental delegations, as in the case of the code regulating marketing campaigns for breast milk substitutes, proposed by the World Health Organization and approved unanimously except for the U.S. vote, the question of the absence of a constraining legal force was raised. Moreover, transnational companies soon found a means of evasion by promulgating their own codes of self-regulation.

The Challenge of Telematics

Third world countries with the resources and the desire were able to endow themselves with the means to implement earnest technology transfer policies. That was the case, in particular, of Brazil and India, which deliberately engaged in import substitution with the aim of developing their own computer technology, aerospace, and weapons industries. They limited the conditions for access to their markets by the largest foreign manufacturers, but at the same time formed alliances with firms that agreed to negotiate the transfer of their know-how, thus promoting national technological independence.

The question of national sovereignty in relation to new information and communication technologies was also troubling the authorities in the large industrialized countries. To meet the challenge, the governments of Japan, Australia, and Canada commissioned their experts to diagnose the situation. In France, an official report on the computerization of society, submitted to President Giscard d'Estaing by Simon Nora and Alain Minc in 1978, was to have international repercussions. It openly argued for a policy of national independence through the reappropriation of the networks of telematics, a neologism they forged to signify the convergence of different technologies in computer systems. According to Nora and Minc, this independence was already threatened by the monopolization of information by data banks belonging to a single power. "Knowledge," they warned, "will end up being shaped, as it always has, by stocks of information. Leaving it up to others, that is to say, American data banks, to organize this 'collective memory,' and being content merely to draw from it, amounts to accepting cultural alienation. The creation of data banks is therefore a condition of sovereignty." Having sounded this warning, the authors then returned to a redemptive view of networks, seen as the guarantors of a renewal of grassroots democracy. "Computerized discussion and its codes," they wrote, "should create a new information agora, broadened to the dimensions of the modern nation."

France set up a commission on transboundary data flows, which warned that "the main challenge remains territorial planning on a worldwide scale, and especially the localization of advanced tertiary activities: the brain of the planet." Finally, a report commissioned by Jacques Rigaud (then a senior government advisor and future head of the private radio station RTL), on "foreign cultural relations," concluded that French culture industries lacked adequate private and public strategies in the international market. The explanation lay in France's particular conception of culture and public service.

Toward a Global Society?

The French experts' evaluation of transnational constraints was not shared by everyone. Moreover, for many, the essential problem lay elsewhere, for technological developments had revolutionized ways of seeing the world. In 1968 Canadian Marshall McLuhan and his colleague Quentin Fiore, basing their analyses on the Vietnam War, the first war broadcast live on television, had already placed their wagers for the future on electronic images. In their view, thanks to the power of television to mobilize its audiences' sensorium, the advent of the "global village"—the retrieval of community through the small screen, which they saw as a form of "communism" more authentic than the Soviet variety—was reducing the threat of war to nil, closing the gap between military and civilian life, and enabling all nonindustrialized areas such as China, India, and Africa to take giant strides ahead. McLuhan embraced the concept of "planetarization" as defined in 1938 by the Jesuit theologian Pierre Teilhard de Chardin. Also in the late '60s, the theoretician of management, Peter Drucker, less inclined to seek for utopia through communication, perceived the new phase of integration of the world economy as the definitive entry into the age of the "global shopping center" and the "global factory," which he illustrated with the example of IBM's production networks.

Zbigniew Brzezinski, an American of Polish origin, future leader of the Trilateral Commission and President Carter's future National Security Advisor, preferred to talk of a "global city" because individuals were likely to be thrown into an anonymous environment. In his book on the "technetronic revolution," published in 1969, he stressed the new interdependence engendered by the communications revolution. "Gunboat diplomacy," he said, "was giving way to network diplomacy." Thus, the notion of imperialism was no longer useful for describing the United States' relations with other countries. This was true because the American superpower, unlike its rival, which exuded boredom and scarcity, had become the "first global society in history." As the center from which the technetronic revolution was being propagated, it was the society that communicated more than any other, because 65 percent of all communication in the world originated there. This omnipresence made it the natural vanguard of a global model of modernity, and the vector for patterns of behavior and values with a universal vocation. In the melting pot of this globality, which transcended firmly rooted cultures, distinct national identities, and traditionally firmly entrenched religions, a new global awareness was, in his view, taking shape. Brzezinski was elaborating on the theses, developed in U.S. political science since the 1950s, about the "end" of ideology, politics, social classes, and confrontations.

Yet at the end of the 1970s, the nation-state was being attacked on two fronts: it was accused of being too large to solve small problems of human existence and too small to solve big ones. It was in these terms that Harvard sociologist Daniel Bell, known for his work on the postindustrial society, addressed the participants of the conference on "Informatics and Society," organized in Paris in the autumn of 1979 in the wake of the Nora-Minc report. As a way out of this dual impasse, information and communication networks had become the panacea.

Chapter 6
Globalization: The Networks of the Postnational Economy

Transboundary logics are undermining the institutional foundations of the communication systems of nation-states. By connecting these systems to the standards of worldwide networks, the process of deregulation is bringing about a profound change in the prevailing economic and social order. It was to define this phase in worldwide integration initiated in the 1980s that the notion of globalization was born. Borrowed from English by other languages, like "internationalization" in the late nineteenth century, it is intended to encompass the entire process of unification of the economic field and, by extrapolation, to account for the general state of the world.

Integrated World Capitalism

The Geofinancial Vanguard

Globalization originated in the sphere of financial transactions, where it has shattered the boundaries of national systems. Formerly regulated and partitioned, financial markets are now integrated into a totally fluid global market through generalized interconnection in real time. The financial sphere has imparted its dynamics to an economy dominated by speculative movements of capital in a context of constant overheating. With the expansion of the speculative bubble, the financial function has gained autonomy from the so-called real economy and supplanted industrial production and investment. The slightest hesitation or misstep sends shock waves throughout the world, foreshadowing crises that are inherent to the system in the absence of supranational regulatory mechanisms. As the first sector of the cybereconomy to be integrated, geofinance, with its abstract and deterritorialized spaces, has heralded the general unlinking of worldwide economic organization from the territories in which national sovereignty is based.

The networks of stock market and financial information, which form a logistic system of global transactions, have been redeployed and now communicate in images the monetary flows that crisscross the planet. In 1983 the Dow Jones group launched The Wall Street Journal Television in the United States for the North American continent, and a year later the Asia Business News, based in Singapore. In 1995 this U.S.-based group established a similar service in Europe. Its rival, Reuters, which had likewise grasped the importance of television when it bought out Visnews, accomplished its 150-year-old financial ambition. Listed on the stock exchange since 1984, ten years later Reuters Holding derived 93 percent of its turnover from the dissemination of economic information (foreign exchange markets, financial futures markets, and stock markets). The news agency thus became the main supplier of electronic traders. Could there possibly be a better image of the *perpetuum mobile* (perpetual motion machine) of media flows on a global scale than their coexistence, in the same enterprise, with constant and rapid flows of dematerialized money?

A Corporate Philosophy

Globalization is above all a model of corporate management that, in response to the growing complexity of the competitive environment, creates and promotes competencies on a global level with a view to maximizing profits and consolidating market shares. It is, in a sense, a framework for interpreting the world that is peculiar to management and marketing specialists. One watchword guides this corporate logic: integration, which implies a cybernetic view of the functional organization of large economic units. In English the word global is synonymous with "holistic." Unlike the French word *mondialisation* (literally, "worldization") and its equivalents in the various Romance languages, which are limited to the geographical dimension of the process, "globalization" relates explicitly to a holistic philosophy, that is, to the idea of a totalizing or systemic unit. The global firm is an organic structure in which each part is supposed to serve the whole. Any shortcoming in the interoperability between the parts, any lack of free interaction, is a threat to the system. Communication must therefore be omnipresent.

Integration here refers to planning, design, production, and commercialization. The total involvement of the employee turned into her/his own manager and marketer and the promotion of the consumer to the ranks of "pro-sumer" or "co-producer" are two noteworthy consequences. But there is also, above all, graduated integration, which heralds a new mode of connection with the world-space. The information and production networks on which the organization of the internal and external circulation of the global firm are based make it a "network-firm." Under Fordism the hierarchical distribution of jobs and authority in a firm corresponded to a sedimentation of geographical spaces; the local, the national, and the international were seen as mutually impervious levels, whereas the new relational conception of the firm, and the world in which it operates as a network, implies interaction between these three levels. Any strategy of the networked firm must be both global and local. It is this permanent interface that Japanese management theoreticians express with the neologism "glocalize," a contraction of "global" and "local."

The objective of dual—internal and external—integration makes the use of symbolic management techniques essential, whether they are called corporate culture or marketing.

Standardization/Segmentation

Unless one embraces the idea that globalization means universal standardization and its corollary standardization of world needs, an extreme hypothesis put forward in 1983 by Theodore Levitt, editor of the *Harvard Business Review,* then the globalization of markets, productive systems, and technical systems goes hand in hand with segmentation. These are the two terms of a dialectic relationship. Massification alternates with demassification. The latter, moreover, helps to push back the boundaries of the former and to overcome resistance to universal standardization. Even firms classified as ethnocentric apply this principle and position themselves in international marketing in such a way as to leave subsidiaries with room to maneuver.

A unified approach at a strategic level is combined with a tactical autonomy capable of catering to all the peculiarities of specific contexts and territories. On the one hand, the adaptability of production tools to particular needs, owing to flexible technologies, makes it possible to produce smaller series of products and therefore to differentiate them and monitor their increasingly reduced life cycles. On the other hand, the cultural brakes on the performance of firms are taken into account by managers who do not dissociate the globalization trend from the conditions of its insertion into a national or local context. Specialists of intercultural communication applied to management have introduced into their taxonomy the notion of intermingling, which denotes the need to avoid head-on collision of cultures within the global firm. Finally, marketing and advertising adjust their intervention to suit the different scales of segmented markets and targeted publics to better take advantage of opportunities for penetrating networks with their products and services. Optimizing investments in advertising leads to the narrowing down of targets. Proof of this is found in the application of new technologies that manipulate image and create "virtual reality." An image-processing software package makes it possible to replace a billboard situated on the site of a sporting event by others, seen only by TV audiences in particular countries or regions. Even before deontological rules had been established, the advertising industry had become an experimental field for new technologies. Segmentation of consumption progresses in step with the improvement of data banks and other computerized techniques for the socioeconomic mapping of targets.

The rise of electronic commerce and the economic model of "mass individualization" that this model theoretically presupposes, requires the enterprise to follow closely the behavior of consumers and reconstitute the catalog of their purchases in order to identify the moment when, for reasons to be determined, they ceased being loyal to a given product and buy another. This new pressure to convert lookers into loyal repeat buyers through one-to-one marketing, with the use of databases and the Internet, creates risk for privacy, and these have not escaped the regulatory authorities of the European Union. Since October 1998, after six years of negotiations, a Directive on Data Protection guarantees European citizens absolute control over data concerning them, to the great chagrin of global free marketers in the United States. Thanks to this directive, no company may transmit personal data about European Union citizens to countries whose privacy laws do not meet European standards; companies must show customers their complete data profiles on request; cross-marketing without customer permission is banned; Web-site owners are not authorized to use data tags or "cookies" to track customers' buying habits and tastes without their consent.

The New Status of Communication

In the transition toward the global managerial model, the proliferation of risks has promoted the function of communication to the rank of a strategic management tool. The need for high visibility has transformed the firm into a political actor directly involved in the management of the community. The symbolic management of various publics by corporations has become professionalized, and communication careers have diversified. The former function of public relations has expanded to become "public affairs," a term that U.S. corporations in the sector had already adopted by the end of the 1970s—a time when they were under attack from all sides—to better mark the entry of the firm into the political-strategic field.

Restructuring, mergers, massive layoffs or "decruitments" (to quote a euphemism inaugurated by communication agencies specialized in such operations), social conflict, ecological disasters (discharges of toxic waste into rivers, supertanker wrecks, accidents in nuclear plants, chemical explosions), terrorist threats, and accidents and incidents of all kinds have been sources of tension demanding an immediate response. So-called crisis communication has attempted to offer a solution by proposing preventive techniques such as surveillance and social observation, as well as dialogue and negotiation with staff, shareholders, customers, the public at large, legislators, and governments. All these extreme situations or, in more technical terms, major technological hazards, have forced firms to think about crisis management. A crisis is no longer only that ultrasensitive period when the firm has to set up its emergency committee and mobilize its resources to cope with a dysfunctional event that disrupts its internal and external relations; it is now internalized by management and inspires a continuous mode of organizing the firm and its communication apparatus under "normal" circumstances.

The managerial model of communication and construction of a corporate image has imposed itself throughout society as the only way of communicating. Communication conceived in this way is henceforth considered an excellent social management technique. Witness its spread into state institutions, local authorities, and humanitarian associations, which define their relationships with citizens or civil society by calling on the know-how and imaginary of marketing.

Economic Intelligence

The shifting and unpredictable framework of the globalization of commercial exchanges has transformed the role of economic and technological intelligence in defining the strategies of corporations and state and para-state actors. The mission of competitive intelligence is to help identify competitive threats from the outside. One clear sign of the growing importance of the collection and systematic interpretation of all data likely to shed light on the behavior of private and public actors can be found in the reorientation of state intelligence services after the Berlin Wall fell. Although economic intelligence is defined in textbooks as the set of coordinated actions concerning the search for, processing, distribution, and protection of information useful to economic actors and legally obtained, in reality, illegal activities often provide the key to interpreting the stock of public information (conferences, seminars, publications, data banks, etc.). A clear example was the CIA's attempts to bribe senior officials in the French telecommunications and audiovisual industry in order to gain information about their strategy at the time of the 1992–1993 GATT (General Agreement on Trade and Tariffs) negotiations.

The latest version of the French penal code, applicable since 1994, shows that today, in the context of exacerbated international competition, the risk of aggression is economic rather than military. Attacks on the fundamental interests of the nation now encompass the essential elements of the economic and scientific potential, hence the recommendations made in 1994 by the ministry of higher education and research, in the form of a guidebook for researchers, titled *Protection of Scientific and Technological Creation and the Vulnerability of Information*. The aim of this guide was to prevent an "illicit appropriation of information on sensitive subjects."

In 1998 the European parliament published, at the initiative of the British representative Glynn Ford, a report titled "Assessment of Techniques for Political Control." This document describes how the United States and its Commonwealth allies (Britain, Canada, New Zealand, and Australia), united in an institution called Ukusa, intercept, with total impunity, telephone calls, faxes, and e-mail throughout the world by means of a network of spy-satellite surveillance code-named Echelon. Originally designed in 1948 by the U.S. National Security Agency (NSA) for political and military espionage in the communist bloc, this powerful and highly sophisticated electronic information network (Elint) became a weapon in the economic war after the Berlin Wall fell. Each country in the Ukusa pact (except Canada) has a specific area of the globe to cover, and the information retrieved is automatically transmitted to the United States. The most important targets include not only major European corporations likely to win contracts at the expense of U.S. rivals, but also humanitarian organizations such as Greenpeace and Amnesty International. The Civic Liberties Commission of the European Parliament considers this problem serious enough to have called for an inquiry into this system and its repercussions in Europe.

The Legitimacy of Expertise

The key position that corporations have come to occupy has changed the balance of power between research for operational or administrative purposes, and research that takes distance from its object (without, however, being isolated in an ivory tower). Here too, in the harnessing of the breeding grounds of brain power, hitherto marginal to the valorization of capital, the fate of integration is in the balance. The mobilization of energies around competitiveness precipitates the encounter between the places in which knowledge is traditionally produced and diffused, such as universities, and the needs of economic actors—especially where the intellectual classes have historically been structured around a critique of social institutions. Synergies that until very recently seemed like a marriage between the incompatible now occur; their aim is to enroll geography, history, ethnology, psychoanalysis, sociology, and linguistics in the service of corporate performance. The problem is not so much the alliance as the terms of the exchange. The risk in this contractualization of human science research is in legitimizing a forceful comeback of the multiple forms of empiricism. Whereas the decision makers, the World Business Class, think in terms of a whole, the "integrated intellectuals"—to quote the expression made fashionable in the 1960s by Umberto Eco who contrasted them with "apocalyptic" or critical intellectuals—are riveted to functional observations at the request of those who commission their research. These observations are atomized and decontextualized in relation to the implications of change in the social and economic model. Relayed by a vast transnational network of public and private training programs in management science, best-sellers on managerial reengineering, or the "third-wave" postcapitalist society, workshops, lobbies, and corporatist organizations, the global business community is becoming the new world elite and it continuously strives to naturalize notions used by all and sundry to label phenomena.

The Quest for the Single-Image Market

Communication Groups and Networks

The creation of a single-image market is an important stake in the quest for a so-called global culture. The forging of single macroregional markets had hardly been announced when communication groups and international TV channels (such as CNN) or regional ones (pan-American, pan-Arabic, pan-Asian, or pan-European) inaugurated the search for cultural universals. Helped along by the integration of communication operations, the third generation of advertising networks, so-called global networks, followed suit, responding to the trend toward the interconnection of markets. One of the axioms in the search for a global common denominator is the cultural convergence of consumers, the product of years of influence of mass culture on the collective imaginary of consumers of diverse cultures. As natural mediums for universality, the culture industries in the United States still appear to be playing an excessively predominant role in defining the parameters of globality.

The construction of these global communication groups and networks required a radical deregulation of national communication configurations, and this has affected both systems in the public sector and those managed according to commercial criteria. Although these groups and networks remain, to a large extent, based in the major industrial countries, other actors have appeared in the audiovisual market. The two classic examples are the Brazilian group Globo (whose name is most appropriate) and the Mexican group Televisa whose series and *telenovelas* have been disseminated far beyond their home continent. Throughout most of the world, emerging markets and secondary markets in the audiovisual sector have appeared. The major event, without any doubt, is the incorporation of the large urban areas of China and India into a zone covered by satellite through the intervention of global communication groups.

The first wave of integration through mergers and acquisitions in the 1980s was followed by a second one in the following decade, stimulated by the promises of digitization as symbolized by information highways. Because any product translated into digital language could circulate in any medium, there was a convergence in the United States between cable operators and film studios, telephone companies and communication groups. After the first stage of deregulation of audiovisual systems, which in 1982 liberalized the rules governing the concentration of channels and stations, Congress renewed the process by removing barriers between cable and telephone companies and between program producers (cinema and television) and distributors. Networks had, prior to this, been prohibited from producing their own fiction and variety programs. The urge to group together as many film and program industries as possible, so as to provide ample material for the hundreds of channels emerging in each country, fostered the alliance between studios and networks (Disney and ABC) as well as mega-mergers between communication groups (Time-Warner-Turner). However, the quest for industrial synergy did not always have the expected results and the course of mergers and acquisitions was strewn with strategic blunders. A case in point is the unsuccessful merger between Matsushita and Universal, in which the U.S. public, in the grip of nippophobia, feared the abduction of the American spirit.

One sign of the process of integration under way are the neologisms that have appeared in technical vocabulary: *advertorials* (contraction of advertising and editorials), *infomercials* (information and commercials), *infotainment* (information and entertainment), and *edutainment*. This hybridization of words reflects the hybridization of information and communication technologies made possible by computers.

From "Television without Frontiers" to the Cultural Exception

While most governments of nonaligned countries abandoned their attitude of protest and rallied to the neoliberal approach, the deregulation of flows of cultural products was met by institutional retaliation from the European Union. This voluntarist strategy was based on one obvious fact: the trade deficit of Europe's audiovisual sector, which had become the most important solvent market of the United States' film, television, and video industries. Year after year, under the effects of deregulation of the audiovisual systems, television and video deepened the European deficit. According to the annual report of the European Audiovisual Observatory, approximately 69 percent of all fiction programs (series, television films, and films) imported in 1994–1995 by eighty-eight television channels in the European Union were of U.S. origin. As far as the commercial exploitation of films was concerned, foreign markets had become increasingly vital for the

U.S. majors, who received over 70 percent of the earnings of cinemas in Europe. Even France, the only country to have maintained a substantial share of the market for its own films, had to face an increasing share of American films shown in its cinemas: from 30.1 percent in 1982 to 57 percent in 1993.

The first preparations for a common strategy were made in 1989. The twelve member countries of the European Community approved the final draft of a directive on "television without frontiers," on which work had begun five years earlier. Article 4 urged member countries to reserve the majority of broadcasting time, "whenever possible," for European productions (fiction films and documentaries). Four years later a set of measures designed to organize the European audiovisual industry, known as the Media Plan, was approved.

In 1993 the debate within Europe spread worldwide with the GATT negotiations. Classified as a service, communication generated direct confrontation between the European Union and the United States. The so-called cultural exception argument of the French government opposed the extension of the liberal rules of international trade to audiovisual productions for the same reasons that it opposed their extension to the public health, environment, or internal security of a state. There were several reasons for France's deep involvement in this question. First, its long tradition of defense of its film industry was rooted as much in a particular conception of culture, the role of authors, and the importance of state intervention, as in the consciousness of the many professional organizations in a country that produces an average of 100 to 120 full-length films per year, and in which the film industry provides approximately 70,000 jobs. Secondly, the French state feared its system for disseminating French cultural influence would lose even more weight in Europe and the world. While professionals in the sector, particularly author-director-producer organizations, were in the forefront of this French mobilization against the project, French communication groups of European or international dimensions, clearly opposed to any form of quota policy, conspicuously refused to take a stand. Yet the argument of cultural exception had a historical precedent. During discussions on the U.S.-Canada Free Trade Agreement, Washington had been forced to concede to Ottawa the right to protect the Canadian cultural identity. Known as the cultural exemption clause, Article 2005 of the Agreement covers cinema, radio broadcasting, sound recording, and publishing.

The application of free trade rules proposed by the GATT implied the abolition of the various systems set up by Europe and each European country to protect its own audiovisual space. Funds to support the film industry at the national and European level, and quotas for the broadcasting of European fiction on television were condemned, in this perspective, to disappear in the long term, in the name of free competition in a free market.

The wrestling match with the GATT ended in December 1993 with recognition of the principle of cultural exception. Hailed as a victory, it was probably little more than a reprieve. Hollywood, the U.S. Congress, and the White House adopted a pragmatic attitude. While consolidating their lobbying of the World Trade Organization (successor to the GATT) and national governments, they avoided all so-called philosophical debate. They wagered on digital compression, which made it possible to greatly enhance satellites' retransmission capacities and thereby spill across national boundaries.

Undeniably, the U.S. State Department was quick to learn from the Euro-American confrontation. In April 1995 a memorandum entitled "U.S. Global Audiovisual Strategy" drew the main lines of its policy for the coming years:

1. Avoid reinforcing restrictive measures and prevent current measures for television and cinema from being extended to the new communication services.
2. Avoid useless dramatization by avoiding debates on cultural identity and try rather to find areas of common interest.
3. Try to take advantage of negotiations on the deregulation of new communications and telecommunications services to destabilize the rule of cultural exception in the audiovisual sector.
4. Prevent the audiovisual policy decided by the European Union from spreading further and, more particularly, prevent it from being taken as a model by former Soviet bloc countries where U.S. investors occupy dominant positions in the audiovisual sector.

5. Multiply partnerships and investments of U.S. firms in Europe.
6. Discreetly strive for adherence to U.S. positions by European operators affected by quotas and regulations (private TV channels, advertising industry, telecom operators).
7. Improve the conditions of investment for U.S. firms by liberalizing existing regulations.

The last recommendation is of a more general nature and is aimed at using all international negotiations in which the extension of free trade is in the balance. This was the case of discussions on two agreements, which provoked serious reservations among countries in the European Union: the New Transatlantic Marketplace Agreement (NTMA), aimed at creating a European Union/USA free trade area based on the model of the North American Free Trade Agreement (NAFTA), and the Multilateral Agreement on Investment (MAI), negotiated in the framework of the OECD, an organization of twenty-nine countries with advanced economies, which poses as a principle the total openness of markets and defends the equal treatment of investors in member countries.

During the negotiations on the cultural exception, other European countries' reactions to the French position were far from unanimous. Most remained in favor of a minimum solution, convinced as they were that fixing quotas would prove in time to have little effect. They expressed this idea once again in November 1995 by voting for the status quo during the debate on the renewal of the Television without Frontiers directive. Three months later the European Parliament took the opposite stand by voting for strengthened quotas. The decision was to be reversed, however, by the end of that year.

Information Highways

The problem of the dependence of the image industries soon combined with that of the new information networks and multimedia services. Digital networks opened up images, so they were no longer confined to the leisure industries, and projected them to the very heart of the reorganization of modes of production and distribution of human societies. The massive construction of telematic infrastructures also involved the question of cultural sovereignty. The problem posed was that of developing an industry that was strong enough to ensure that new networks and services did not provide software produced only by the multimedia giants, and that culture industries did not remain chronically in deficit. Another stage in the Euro-American disagreement was thus played out.

In February 1993 President Clinton's administration announced the Gore Plan for the construction of information superhighways. At the end of the same year, the White Paper prepared by Jacques Delors, then president of the European Union, launched the European inforoute project. Growth, competitiveness, and employment were the three leitmotivs of this program, which called for a mobilization of the entire European industrial apparatus. In 1994 a group composed of twenty major industrial firms, presided over by Commissioner Martin Bangemann, submitted a report entitled "Europe and the Global Information Society," which set out the strategic and financial aspects of the project.

In anticipation of the new technological challenges posed by the proliferation of broadcasting channels, the White Paper was followed by a Green Paper on strategic options for strengthening the program industry in the context of the European Union's audiovisual policy. The aim was to build a regulatory framework and a credible financial base in order to limit the fragmentation of European audiovisual markets and enterprises. This would allow for a better utilization of the potentialities of the digital revolution that was transforming the European market into a major stake in all struggles in the world market. A further aim was to try to turn to Europe's advantage the cultural diversity of the members of the Union, which had hitherto been seen as a handicap. On the horizon, directly related to the concerns formulated in the White Paper, was the promise to create, within five years, two to four million jobs in a Europe that, at the time the report was written, had 18 million jobless. Although many economists contested this figure, one thing was certain: the specter of depression put the employment argument at the heart of controversy over the defense of identities.

In February 1995 the G7, the group of the seven most industrialized countries, met in Brussels for a summit on the new information and communication technologies. The United States was represented by its vice president. Invited for the first time to a meeting of this kind, forty-five businessmen from the United States, Europe, and Japan emphasized the urgent need to speed up the deregulation of telecommunications services and the abolition of public monopolies so as to accelerate the deployment of the future electronic highways. The deliberations of the G7 and the recommendations of the Bangemann report concurred in emphasizing that private initiative had to be the driving force of the information society. In the short term, the removal of all obstacles to free competition would lead to the liberalization of telephone networks.

The title of Albert Gore's speech in Brussels was programmatic: "Toward a Global Information Infrastructure: The Promise of a New World Information Order." In July 1997 President Clinton confirmed a resolutely neoliberal approach to the Internet when he proposed making electronic trade a self-regulated global free trade area.

Things sped up after that historic decision. In December 1997 Martin Bangemann, the European Union commissioner in charge of telecommunications, published the *Green Paper on the Convergence of the Telecommunications, Media and Information Technology Sectors, and the Implications for Regulation towards an Information Society Approach.* In the name of this technological convergence, he suggested combining the regulations applicable to the audiovisual sector and those in force in the telecommunications sector and subjecting both to a simplified method dictated by market forces. In fact, following these recommendations would amount to aligning European policy with the desiderata expressed in the document "U.S. Global Audiovisual Strategy" mentioned above. The obsession was to do away with special treatment for products of the intellect, that is to say, in the name of technological evolution, to consider a television program or a film as equal to a telephone call.

Retaliation was immediate. In January 1998 the French regulatory body, the Conseil supérieur de l'audiovisuel (the High Council on the Audiovisual Media), leveled fierce criticism at the Green Paper, summarized in the following five main points:

1. It is advisable to remain moderate and cautious as regards the extent, pace, and consequences of convergence, particularly as long as the audiovisual equipment of the general public is differentiated in relation to the services offered.
2. Firm regulations remain necessary for guaranteeing market equilibrium and the preservation of the general interest.
3. The audiovisual sector must remain subject to specific regulations, considering the stakes involved in its activities, the most important of which is freedom of expression.
4. The determination of the legal framework applicable to services must remain based on the nature of the act, over and above the technical modalities of its execution.
5. The existence of a single regulatory authority for the telecommunications and audiovisual sectors is a secondary question compared to that of the legal frameworks applicable to these sectors. In any case, this question is a matter for nations to determine and, for the moment, seems premature.

Although it expressed little more than a proposal, the *Green Paper on Convergence* showed how sharply positions continued to differ, even within the European Union itself, regarding the path to take toward the global information society. The current negotiations on the General Agreement on Trade in Services (GATS), the WTO's "Millenium Round," will be a test of European cohesion.

"Freedom of Commercial Speech"

From the European Union to the GATT, from the WTO to the G7, the new global framework of managerial thinking shifted the center of gravity of international negotiations on the flow of immaterial data. This shift concealed another: the shift of the very definition of freedom of speech. Citizens' freedom of expression was put into direct competition with the freedom of commercial speech, presented as a new human right. Constant tension can be observed between the absolute sovereignty of the consumer on the one hand, and the will of citizens, guaranteed by democratic institutions, on the other. This demand for freedom of commercial expression was the key issue around which interprofessional organizations (advertisers, agencies, and media) organized their lobbying efforts, starting with the first skirmishes over television without borders. The aim of this demand, which became a leitmotiv as the debates progressed, was to push back the limits imposed by civil society on the "subjection of the public sphere to the ends of advertising," in the words of the German philosopher Jürgen Habermas.

The main idea was the need to allow for the free play of competition in a free market between individuals with a free choice. It can be expressed roughly in the following terms: "Let people watch what they want to. Let them be free to judge. Trust in their judgment. The only sanction applied to a cultural product should be its success or failure on the market." Between that and legitimizing the cultural subordination of certain peoples and cultures—what until the late 1970s was called cultural imperialism—there is only a small step. This step is particularly easy to take because this idea makes common cause with another one: freedom of commercial expression, the new ordering principle of the world, is indissociable from the old principle of the free flow of information. Moreover, by recycling this notion, the doctrine of globalization relegitimizes, in the name of fluidity in the information era, the strictly economistic, Anglo-Saxon conception of copyright that automatically grants all rights to the producer while devolving, more than ever, the validity of the moral right of the author.

In the first case, artists sell their rights, often for a set amount, to producers who are free to use them however they wish. In the latter case, creators remain the sole masters of their work and royalties are paid to them or their beneficiaries until seventy years after their death. Defenders of full and inalienable intellectual property rights refer to the discomfort of American creators and cite such extreme cases as that of composer Cole Porter, who died destitute after selling his rights for a pittance. There is no doubt that in the successive modifications of the notion of intellectual and artistic property rights as defined in the Bern convention, signed in 1886 and revised in 1971 for cinema, in 1991 for computer software, and in December 1996 for multimedia, a major objective has been the free market globalization of culture and information industries. Many see this as the key stake in the worldwide economic war in the area of culture.

From Democracy to the "Global Democratic Marketplace"

Messianic discourses about the democratic virtues of technology, which mask what is at stake in the struggles for control of the structure and content of knowledge networks, are of use in geopolitics. The champion of information superhighways, Albert Gore, adopts the same tone as the prophets who have preceded him since the end of the eighteenth century, when he presents to the "great human family" his world project for a network of networks: the *global information infrastructure* (GII). Addressing the delegates of the International Telecommunications Union at the first World Telecommunications Development Conference in Buenos Aires in March 1994, the U.S. vice president declared: "The final and most important principle is to ensure universal service so that the global information infrastructure is available to all members of our societies. Our goal is a kind of global conversation in which everyone who wants can have his or her say.... The GII will not only be a metaphor for a functioning democracy; it will in fact promote the functioning of democracy by greatly enhancing the participation of citizens in decision making. And it will greatly promote the ability of nations to cooperate with each other. I see a new Athenian Age of Democracy forged in the fora the GII will create.... Let us work to link the people of the world. Let us create this new path as we walk it together." According to Gore, networks used by private actors are tools of development and solidarity and should make it possible to solve the major social and economic imbalances burdening our planet.

History has come full circle. A little more than two centuries ago, the notion of communication entered into modernity via the development of roads. The advent of the postmodern age of immaterial networks and intangible flows is occurring under the sign of the metaphor of highway networks, echoing the collective memory of the great public works projects that favored a dynamic of economic growth in the United States in the 1950s.

On the Global War

Information Dominance

Economic competition caused military metaphors to flourish in the 1980s. More than one treatise on global management explicitly drew upon the works of Sun Tse or Karl von Clausewitz for lessons in waging the economic war. While reducing these market discourses to their true proportions, the Gulf War (1990–1991) also demonstrated the permanence of military logics in a communication environment that the previous decade had induced people to theorize exclusively in terms of geoeconomy. This extreme situation served to reveal the opaque aspects of the information systems that we tend to forget in peacetime, thanks in part to the ideology of transparency from which the new communication society claims to draw its inspiration.

The Gulf War was a war of communication for two reasons. First, the Pentagon developed careful strategies of information and censorship. In particular, it organized pools of hand-picked journalists who went into the field accompanied by a so-called Public Affairs Officer who chose and prepared the troops to be interviewed, controlled television filming, examined photos, and doctored written reports, deleting all information considered sensitive. The Gulf War, and specifically Operation Desert Storm, launched on January 17, 1991, was, in a sense, the armed forces' revenge. Experts in psychological warfare had learned their lessons from the Vietnam War. In 1982, during the Falklands War, the British army acted in a similar way by decreeing an embargo on information. In 1983, during the U.S. Marines' invasion of the island of Grenada, the Pentagon had also shut off all access to the scene of the operations.

Secondly, the Gulf War was waged with information and communication technologies and "intelligent weapons." In the field, behind the "surgical strikes," were missiles piloted by their own onboard computers, reconnaissance satellites, command systems relayed to all the war machinery and even to the weapons themselves and their neural networks. The integration of weapons with multiple networks, "Information Dominance," has its acronym, C4ISR, which stands for Command, Control, Communications, Computing, Intelligence, Surveillance, and Reconnaissance. Moreover, in the background, through its supplies logistics, the Gulf War was the first large-scale war to be run on a just-in-time basis, borrowing the methods of flow management developed by Japanese car manufacturers. (Producers who practice this method stock very little or none of the products they are not sure of selling; as far as possible, they produce only to order. Computers allow orders to be transmitted instantaneously from the distributor to the producer, or even upstream to a subcontractor.)

The nature of war had changed. It was a "global" war, to quote Paul Virilio, not so much in its transhorizon scope and the wide range of its missiles, but in the transfer of responsibilities to industrial and economic programming: shoot and forget.

Such systems remind us, in passing, that well before the language of globalization became popular in the field of geofinance, it had earned its stripes in the armed forces. Communism was the global enemy and technologies for observing that enemy were called global. The Global Positioning System (GPS), in operation from the 1970s, was one such technology. It spawned a constellation of twenty-four satellites providing constant visibility of any point on the globe, as well as digital technology destined to equip the future soldier. Equipped with a GPS receiver, a soldier can be constantly informed of his position, which is shown, along with that of his comrades, on the map projected on the visor of his helmet. Headquarters are also fully aware of their positions and of what they can see because they emit an IFF radio signal (identification, friend or foe). In combat, all the controls of a soldier's weapons appear on his visor. A range finder indicates his distance from the target. Biological sensors even transmit real-time information on his physiological state to his commanders.

Kosovo, or the Dark Side of Globalization

"Where conquest has failed, business can succeed"; "the global corporation as the instrument of world development is the only force for peace." Such declarations, typical of recent economic summits, illustrate how the new global elite has attributed to itself the messianic role of midwife of world peace, in the name of an undefined global social conscience. In a remarkable interview on the French television channel Arte in 1997, Ted Turner, the founder of CNN, the leading planetary channel, expressed this outrageously neomillennialist view: "We have played a positive role. Since the creation of CNN, the Cold War has ended, the conflicts in Central America have come to a halt, and peace has come to South Africa. People see that it is stupid to make war, and nobody wants to look stupid. With CNN, information circulates throughout the world, and no one wants to look like an idiot. So they make peace, because that's smart."

Two years later, the Realpolitik of the allied forces in ex-Yugoslavia exposed the naiveté of this crusading determinism that, since the Berlin wall fell, has been a key ingredient of the sagas of conquest of the global image market, leading some to believe that worldwide television alone can shape and reshape the world. The prophetic discourse, associated with figures such as Peter Drucker and Bill Gates, about the rise of a "frictionless capitalism" operating thanks to a total liberation of the forces of nature, the market, and technology, reveals itself to be no more than a simplistic response to complex questions. The war of Kosovo has thrown a stark light onto military doctrines of global security—the dark side of globalization as the planetarization of free-market democracy. The inner core of this vision—the reshaping of the world through marketization—is expressed, in more aseptic and diplomatic language, as the idea of a peaceable commonwealth of responsible democracies, joined together by commerce and liberal ideals. In the language of strategists, this translates into "globalization cum unipolarity." The concentration of geopolitical power under the hegemony of the United States, "the lonely superpower" in Samuel Huntington's words, is the logical corollary of geoeconomic globalization, defined as "decentralization at the planetary level." This strategic concentration guarantees the free deployment of commodity flows, wherever the "lonely power" and its associates see it as in their interest.

The invoking of the discourse of the "humanitarian war" or "moral war" to legitimize the use of force reflects a significant and disturbing evolution, which enjoins us to take seriously the fear, expressed by critics of the Kosovo expedition, that NATO's "strategic concept" may be drifting toward a role of planetary intervention, necessarily selective of course, to the detriment of the universal mission invested in the United Nations by the community of nation-states for the settlement of conflicts. This expansion of NATO's area of intervention is all the more paradoxical when one considers that the original reason for the alliance's founding—opposing the Soviet threat—has long since disappeared. Such concerns, provoked by this last major war of the twentieth century, contrast sharply with the futile promotional slogans that, since the rise of deregulation, have persistently presented globalization as just another commercial gimmick, as in this announcement for the World Trade Congress in Singapore in 1996: "Globalization has arrived.... Can you navigate the New World Order?"

By calling attention to the opacity, the complexity, and the spatial and temporal depth of the facts, recent controversies over the role of cultural, religious, and political history in the Balkans in the genesis of the conflict have upset the naive and triumphalist discourses of globalization-as-redemption and the end of ideology, class conflict, and politics more generally.

Chapter 7
The Fracture: Toward a Critique of Globalism

"Naming things badly adds to the misfortunes of the world," said Albert Camus. Globalization is one of those tricky words, one of those instrumental notions that, under the effect of market logics and without citizens' being aware of it, have been naturalized to the point of becoming indispensable for establishing communication between peoples of different cultures. This functional language constitutes a ready-made ideology that conceals the disorders of the new world order. Thus, the time has come, in this period of international integration, to distinguish between globalist mythology and concrete reality. Contrary to the economistic vision of a world united by free trade, a rift is appearing between specific social systems and a unified economic field, and between singular cultures and the centralizing forces of "global culture."

A New Map of Inequality

World-Communication: The Tropism of Global Flows

The integration of economies and communication systems spawns new disparities between countries and regions and between social groups; it is to these logics of exclusion that the concept of world-communication refers. Contrary to what the egalitarian and globalist representation of the planet suggests, it serves as a tool for analyzing the globalization of the system in progress, without fetishizing it, by restoring its historical concreteness. It reconnects with the history of world trade and the social and economic disparities accompanying it. Based on Fernand Braudel's concept of world-economy, it reminds us that networks, embedded as they are in the international division of labor, organize space hierarchically and lead to an ever-widening gap between power centers and peripheral loci.

Four major changes have contributed to the creation of new divisions in the world:

1. The sudden appearance of newly industrialized countries, especially the "dragons" (South Korea, Hong Kong, Singapore, and Taiwan) and their emulators in Southeast Asia, before the financial crash in 1997 that affected a large part of Asia and undermined the economies of these countries, held up as new models of development.
2. The construction of the large free-trade economic blocs around the Triad powers (North America, East Asia, and the European Union).
3. The political and military hegemony of the United States, both policeman and mediator, omnipresent in situations of conflict throughout the world and striving, more than ever, to promote its model of capitalism as the only road to a global economy and global society.
4. The decline of the third world as a subject of history.

But although the north/south dividing line no longer suffices to make sense of the world, the structural inequalities criticized in earlier decades have not disappeared. What has disturbed the Manichaean representation of the world is that the north has discovered on its own territory people from the south, and that in the very heart of the south have emerged "norths" with their own "souths."

Everywhere, new forms of competition bring territories into opposition and lead to differentiated uses of them. In the organization of economic space and the struggle for optimum use of different territories, two contradictory tendencies are at work: first, a process of delocation/relocation toward areas with low labor costs (a process in which rotation occurs fairly rapidly, judging by South Korean firms' strategies of expatriation toward China and Southeast Asia in the 1990s, which may well undermine the "dragon" model of national development, independently even of the destabilizing effects of the financial crash in 1997 throughout the region); second, there is a process of metropolization or concentration of investments in innovative regions with diversified skills and high technological density. To the economies of scale or productivity gains resulting from the size of a firm, and to the economies of scope related to the diversity of its activities, have been added economies of agglomeration. The convergence around distinct poles and the organization of the world economy into networks linking these poles—to the detriment of the areas in between that are less well endowed and therefore more exposed to marginalization and abandonment—carry a risk of splitting the world economy in two and creating a two-speed social geography. This is what has been called the "archipelago economy" (P. Veltz) or "global techno-apartheid" (R. Petrella). At the four cardinal points of the planet, and to differing degrees depending on the country and continent, these megalopoli, the centers of world networks and markets, generate tropisms that are reflected in the map of telecommunications networks and flows. In Thailand, Bangkok has 68 percent of all the telephone lines available in the country. In Brazil, the density of information and communication technology in the São Paulo area is similar to that of the European "golden triangle" and light years from that of Recife, even though the São Paulo megalopolis also has its own urban periphery characterized by ruralization, a phenomenon typical of Latin American cities.

Observations such as these are for marketing specialists the foundations of geostrategies of segmentation or creation of "consumption communities." Considering that variations in lifestyles and standards of living are more important than geographical proximity and belonging to a national tradition, the advertising industry strives to construct vast transnational communities of consumers who all share the same "socio-styles," forms of consumption, and cultural practices. In a sense, these typologies of transborder marketing targets merely confirm a structural imbalance: on the one hand, a proliferation of ubiquitous symbols of global culture and, on the other, a declining proportion of beneficiaries of the commodities and the lifestyle these advertising messages put on display.

The reproduction of strong tendencies toward segregation between data-rich and data-poor groups, a risk highlighted even in the most official documents, refers both to these groups' connection to the world information infrastructure and to the compilation of their own stocks of data. Thus, in the 1996 World Bank report, "Increasing Internet Connectivity in Sub-Saharan Africa," concern is expressed about the fact that most countries in this region are not linked to the Internet at all (as compared with the explosive growth observed since 1988 not only in the major industrial countries but also in Eastern Europe, Latin America, and Southeast Asia). The discourse wavers between brilliant expectations from the "information revolution," seen as affording an exceptional opportunity to leap into the future in a clear break with decades of stagnation or decline, and gloomy prospects for those countries unable to take advantage by surfing the great wave of technological change, and therefore likely to be overwhelmed by it.

The fact that, according to the estimate of the MIT experts, on the eve of the third millennium barely 2% of the world population was connected in one way or another to a world information network, requires us to relativize the euphoric celebrations of the power of technology to upset planetary hierarchies. While recognizing the undeniable potential of the Internet to broaden the number of actors in scientific research, we must nonetheless remember the observation made during the world conference on science in the twenty-first century, organized in June 1999 in Budapest by UNESCO and the International Council of Scientific Associations, in the presence of some 2000 participants, including scientists and economic and political decision makers: the North-South science gap is becoming ever wider, and brain drain is only aggravating the situation. As one indication of this increasing concentration of research activities, public as well as private, nearly 85 percent of world investments in research and development are made within the triad powers, whose respective shares are as follows: North America, 37.9 percent; Western Europe, 28 percent; and Japan (plus the newly industrialized countries), 18.6 percent.

Parasitic Networks

When some political scientists speak of the "new global fronts of disorder," "areas of darkness," or "anti-worlds," they are referring to fundamentalism, sects, channels of the underground or informal economy, Mafia networks and illicit trafficking (from narcotics to children or contraband electronics), transnational flows of diasporas and migrant labor—both regular and illegal—toward the affluent countries and regions, etc. These dissonant fronts and parallel worlds reveal the crises, conflicts, and imbalances affecting our changing societies and confront them the constant risk of collapse or disaster. The following examples illustrate the importance of these fragmented fronts of world disorder.

- Specialists of virtual reality point out that the dematerialization of the economy and the generalization of delocalized, virtualized cyberenterprises skilled in eluding all tax and social control by nation-states used to governing a real territory, favor the emergence of virtual tax paradises, "virtual lotteries" and "cyber-casinos." Internet casinos are made available to consumers from a wide range of territories where gambling is authorized, from Las Vegas to various Caribbean islands, Hong Kong, etc. Extraterritoriality also opens new avenues for laundering money.
- Satellite dishes, stationary or mobile radar, and monitors, trackers, and surveillance aircraft that form the basis of the ecological watch system (Sivam) being set up by Brazil in the Amazon basin are not exclusively intended to preserve the balance of the planet's "lungs" and biodiversity. The function of this complex system is also to detect irregular movements of air traffic and to fight crime such as drug traffic.
- By prohibiting individual satellite dishes, the Saudi Arabian and Iranian governments are trying to shield their

citizens from all cultural expression that is contrary to the Mollahs' message. In France, the frequent use of satellite dishes by North African immigrants nurtures fears of an uncontrollable shower of Islamic propaganda in certain Parisian suburbs; as a direct consequence, the French government created an Arabic channel on the cable, so as not to leave the field open for the satellite.

- In the face of growing social exclusion and mounting violence among society's rejects, the logic of security is becoming prevalent on every level: individual and collective, national and international. This imperative of security directly affects the definition of the ways in which new information and communication technologies are introduced into society. The expansion of the market for electronic security devices is clearly the most obvious sign, but there are other, less obvious effects. For example, the cell phone has spread much more quickly among the middle classes of Caracas than in European Union cities. The reason is not only the lack of fixed public telecommunications networks, but also the fact that this technique for signaling one's position has proved to be a valuable tool for personal and family defense in a city with one of the highest levels of delinquency in the world.

The potential for conflict is thrust onto a world where the ecological balance is turning into an issue of survival for humanity and where the specter of a food crisis is emerging. On the verge of collapse due to overuse of arable soil, deforestation, depletion of water resources, and intensive exploitation of the ocean's resources, and no longer able to meet the demands of the world's growing population, local ecosystems have become sources of social crisis. In 1996, in an interview for the newspaper *Le Monde,* Lester R. Brown, head of the leading international ecological research center, the Worldwatch Institute in Washington, noted that there were more conflicts concerning the sea in the year 1995 than throughout the entire nineteenth century. "The war between man and the Earth is already on.... We tend to forget that natural systems are the foundations of the world economy and that whatever the prospects might be of telecommunications and computer technology, if the productivity of ecosystems diminishes, the perspectives of the world economy will deteriorate.... We must, therefore, rethink the structure of our economic system and its values, our models and ways of life."

The Boundaries of Monoculture

Avoiding Manichaeanism

This dilemma has come to pervade reflection on the cultural future of the world, under the impact of the symbolic universals of mass consumption and real-time networks. Some consider that the establishment of a McWorld is inevitable, because monoculture is the price to be paid for free trade and the creation of large economic blocs. In diametrical opposition to this collective representation, others think that cultural homogenization is not the order of the day in a world racked by social and economic fractures and rigid nationalisms. For the latter, the Jihad has been turned into a symbol of the state of the world.

To what extent do these extreme images account for the complexity of the future of culture(s)? How can this historical phase in the evolution of our societies be defined without falling prey to the catchall, facile words and phrases, the proliferating notions of homogenization, standardization, and massification in all their variants?

Vicissitudes of the Global Village

History has often revealed underscored the flaws in the representations centered around the global village that have fed the public imagination regarding the future of the human community and, in the Realpolitik of business corporations, constituted an inexhaustible supply of arguments to justify the grand sagas of conquest of the world market.

While there is no doubt that the importance assumed by the media has radically changed ways of making war and representing it, the worldwide dissemination of conflicts by satellite has certainly not helped to break down the wall between the military and civilians. Nor have the global media helped developing countries to catch up with those leading the way in the industrial world.

The extensive media coverage in 1992 of the humanitarian mission in Somalia, Operation Restore Hope, contrasts with the conspiracy of silence shrouding the sheepish departure of the troops and their hypersophisticated equipment, and gave citizens neither the possibility nor the desire to change things. From Bosnia to Chechnya to Liberia, the habit of seeing shocking images of human barbarity has caused audiences' attention to wane. Humanity is still waiting for the realization of Marshall McLuhan's and Jack London's prophesies regarding social and pacifistic revolution through animated pictures.

The multiplication of so-called global events—those cathartic events that gather the most diverse local and national publics around the same news, reports, and programs—has not proved to necessarily create a closer-knit world community. One is rather tempted to think the contrary when listening to the comments of journalists from various nations during major international sporting events, for here we move in leaps and bounds away from the myth of the global village toward narrow chauvinism—a troubling development considering that the way in which sports competitions are presented serves more and more often as a reference or model for presenting information on war.

We may also express serious doubts about the new versions of the end of ideologies myth, which was revived when the Berlin Wall fell, for their main argument is the globalization of mass culture. Francis Fukuyama recycled it in the form of the "end of history" myth following the events of 1989. The fact that transistors had become commonplace in the People's Republic of China, that Mozart was played as background music in Japanese supermarkets, and that rock music was the expression of a revolt against a dying Stalinist ideology in Prague were, for this high official in the U.S. State Department, an irrefutable sign of the democratic homogenization of the world under the auspices of neoliberalism. Since then, the idea has taken root in free trade rhetoric that the spread of products of the entertainment industry automatically leads to civil and political freedom, as if the status of the consumer were equivalent to that of the citizen.

The Hybridization of Modernity in Question

Since the beginning of exchanges in the world, the cultural and institutional models conveyed by hegemonic powers have encountered peoples and cultures that have either resisted annexation, been contaminated, imitated their conquerors, or disappeared. In these cultural melting pots, syncretisms have been born. For example, by appropriating in an original way the liturgies, modes of representation, or laws imposed on them by the conquistadors of New Spain, native peoples took them out of the grasp of those who had invented them. This reterritorialization is a key element of hybridization (*métissage*) and the Baroque will. One of Michel de Certeau's contributions was to highlight, in his book *The Practice of Everyday Life,* the incessant movement of interaction between a system that is imposed and its users; he extended this analysis to the numerous tactics employed daily by the ordinary individual user and consumer of commodities to subvert the networks of discipline and social control.

While globalization is a component of contemporary culture, it is not the only logic capable of shaping the destiny of the planet. The 1980s, which saw doctrines of financial globalization and cultural standardization flourish, also witnessed the emergence of a current of thinking focused on the disjunction between the centripetal or agglomerating forces of market logic and the plurality of cultures. Fragmentation and globalization are conceived as a couple in tension, in which the decomposition/recomposition of social and cultural identities are played out. New questions are raised: how does linking up to global networks become meaningful for the different communities? How do they resist it, adapt to it, or submit to it? Such questions were already haunting H. G. Wells's forecasts at the dawn of the twentieth century, when he examined linguistic hegemony. New perspectives on the bonds being established between the global and the local have emerged, in a break with the earlier idea of the inevitability of monoculture.

Anthropologists have undertaken a critique of conventional discourse on the relationship between transborder cultural flows and singular cultures. For them, the intensification of the circulation of cultural flows and the very real tendency toward the globalization of culture lead not toward the homogenization of the planet but toward a more and more hybrid world. The notions of hybridization and *métissage* account for these combinations and this recycling of transnational cultural flows by local cultures. Thus, the Indian anthropologist Arjun Appadurai sees the tools of homogenization (weapons, advertising techniques, the hegemony of certain languages, styles of dress, etc.) introduced through globalization, as being "absorbed into local political and cultural economies only to be repatriated as heterogeneous dialogues of national sovereignty, free enterprise, and fundamentalism, in which the state plays an increasingly delicate role." With too much openness to global flows, the nation-state risks revolt (the Chinese syndrome); with too little, the state is banished from the community of nations (the case of North Korea, the world's last autarkic regime). This anthropologist even dares to speak of "alternative modernity" and the "explosion of cultural modernities" that are emerging in Bombay, Tokyo, Rio de Janeiro or Hong Kong, Los Angeles, New York, London or Paris and attest to the diverse singularities' multiple paths of access to new forms of cosmopolitanism. Demystifying the notion of modernization as a univocal projection of the European and North American experience, which dominated references to the question of development and underdevelopment until the 1970s, the new concepts seek to show that local cultures, far from disappearing from the world map, are being reformulated. By mixing the modern with the traditional, they are laying the foundations of their own culture industries and fields of artistic creation. This is clearly shown by various phenomena, including the breakthrough of the Brazilian television industry in the world market, and the vitality of artistic creation in the field of dance in certain Black African countries. But these phenomena are masked by persistent clichés about the poverty of these areas. The rapidity with which Asian and Latin American countries have adapted to digital technology and the advantage they have taken of it, both to perform in the world market and to link up social or scientific research networks, are indicators that our stereotyped images of this large part of the world need to be revised. We cannot deny, however, that these new sources of modernity coexist—as the second side of the coin—with a galloping process of impoverishment and exclusion of large sections of the population. The new hypotheses on intercultural relations indicate that, throughout most of the world, a process of reviving and celebrating particular cultures has begun and that it is a precondition for the invention of an economic and social model no longer totally subjected to the demands of foreign markets.

An Anthropology of Contemporaneousness

The crisis of the ideology of progress/modernity is also haunting the work of anthropologists in the major industrial societies. Ways of looking at others have changed, for the crisis of social meaning (the instituted and symbolized signification of one's relations with others) has become generalized throughout the world. Every individual has become aware of belonging to a planetary society. All are contemporaries, but within a plurality. How can the unity of the planet and the diversity of the worlds constituting it be conceived together? That is the question underlying new anthropological perspectives on the world's complexity. The focus has moved away from distant, "exotic" societies and cultures, the subjects of classical social anthropology, and shifted toward a search for contemporaneousness. This is a response to the acceleration of history and changes of scale, the shrinking of the planet and the individualization of trajectories and references. Here we may quote Marc Augé, whose evolution in research—from the ethnology of African societies to that of Euro-Disney, the subway, and megalopoli—is representative of the approach that observes the new modalities of symbolization at work throughout the world as a whole: "These modalities involve information networks that are the instruments par excellence of extended mechanisms of ritual behavior, the particular elaborations of individuals who are in varying degrees integrated into these networks, and a whole set of official and unofficial institutions that are all trying to construct significations of compromise between networks and individuals. From this point of view, an African prophet-healer, a group of architects working on a development project, or a medical team discussing the form of its intervention in a particular environment, all constitute realities of the same nature." He concludes his argument as follows: "Adapting to changes of scale does not mean that one no longer privileges the observation of small units, but that one takes into consideration the worlds that cross through them and extend beyond them and, in so doing, constantly constitute and reconstitute them."

The anthropologist proposes two dichotomies for describing the contemporary world and spaces: place versus nonplace, and modernity versus "supermodernity." Place is triply symbolic for it relates to identity, relationships, and history. It symbolizes the relationship of each of its occupants with her/himself, with the other occupants, and with their common history. The multiplication of "nonplaces" is a characteristic of the contemporary world: places of traffic (highways, flight corridors), consumption (large supermarkets), and communication (telephone, fax, television, networks). In these nonplaces people coexist and cohabit without living together. The status of the consumer or lone passenger involves a contractual relationship with society. These empirical nonplaces that generate mental attitudes and types of relationship with the world belong to supermodernity, defined by contrast with modernity. Because today's metropolis is presented as a central point, a node of relations, transmission, and reception in the vast network that the planet has become, it is at the crossroads between places of modernity and nonplaces of supermodernity.

The merit of this ethnology of industrial societies is that it provokes a radical questioning of the ethnocentric perspective on others. Because of this shift, an inversion of perspectives becomes possible: the center is seen from the periphery and the different regions of the north are seen from those of the south.

Theory Put to the Test of Free Trade

Interest in the singularity of cultures has also led to new ways of considering processes of individual and collective reception of transnational cultural goods in the various milieus in which they are consumed. The reception of products that are emblematic of media globalization, such as the television series *Dallas* and *Dynasty,* have thus been the focus of particular attention. Theories of behaviorist inspiration on the univocal "effects" of television programs have been replaced by other explanations that emphasize mediations in the construction of meaning and highlight differentiated receptions, according to the culture to which the addressee belongs.

The recognition, which is tending to become generalized, of milieus and mediation, is at the origin of a revival of theoretical perspectives regarding the analysis of relations between cultures. But some have seized this opening as a pretext to claim the end of the domination of certain cultures by others, and to proclaim the advent of a sovereign consumer who moves around in the sphere of global culture with no limits other than her/his own free will. Power relations and socioeconomic determinations are thus expelled from the field of cultural analysis, and all possibility of political understanding of the world evaporates, giving way to cultural hyperrelativism. The new concepts are made to play the role of a smoke screen because they are associated with all sorts of compromises with the social and productive order. Thus, if these new conceptual tools are to retain their heuristic force, they need to be handled with the epistemological caution required by a period in which the spirit of free trade has pervaded the social sciences. Certain ethnologists' critical questioning about the operational use that can be made of their approaches to contemporaneousness serves to recall the ambiguous status of these changes in perspective in the analysis of everyday use and users of places and nonplaces.

The emphasis placed on mediations and interactions must not cause us to forget that contemporary universalization of a productive and technoscientific system remains, more than ever, marked by the inequality of exchanges. Moreover, the powerful thrust pushing peoples and nations to reappropriate their history and culture is highly ambivalent. It could lead to the temptation of the extreme, i.e., closure within a single identity, the return of tribalism, the purity of ethnocultural identities, and multiple expressions of intolerance toward foreigners. Can these radical forms of exclusion of the Other also be interpreted as a confused retaliation to the exclusion inherent in the segregationist logics of economic globalization?

A Clash of Civilizations?

No part of the world is safe from the ethnocultural temptation that, suddenly or brutally, crisis introduces into practice and theory. With the end of the Cold War, new Manichaean visions of the world are emerging, based on the idea of civilization interpreted as a community of history, language, culture, tradition, and, above all, religion. Such visions play on the idea that it is necessary to build fortresses as a defense against the Other. We see an example of this in the digression of a certain strand of geopolitical thinking that extrapolates from the lessons of local conflicts with a significant cultural dimension, such as the Bosnian and Caucasian wars, and defines the sense of belonging to a civilization as the main motive for wars in the third millennium. That is the argument developed by Samuel P. Huntington, who considers that the origins of war will henceforth no longer be economic and ideological, but the defense of the boundaries of civilization. Seven or eight main civilizations will confront one another in the future: Western, Confucian, Japanese, Islamic, Hindu, Orthodox-Slavic, and, perhaps, African. By focusing on "Muslim militarism" and seeing in a potential alliance between the Islamic and Confucian civilizations the birth of a pole that is potentially hostile to the West or defying the power and values of the West, Huntington is identifying the "new global enemy" and urging countries belonging to Western civilization to draw conclusions for their own security. Global policy must not only strengthen bonds between countries and groups that defend the same values and interests, but reinforce the "maintenance of American military superiority in East and Southwest Asia."

This culture-based and security-oriented forecasting not only disregards the structural sources of economic and social exclusion in the world, but also overlooks the fact that the threat in the era of growing international pressure on local and national realities is not one of civil war between civilizations but rather one of confrontation within each culture. This is where the fate of the globalization/fragmentation dialectic will be decided.

Toward a New Democratic Cosmopolitanism

Global Causes

Nongovernmental organizations (NGOs) have suddenly appeared in the forefront of the international scene. Hitherto invisible, their numbers have multiplied, and their protagonists, networks, and social purposes have diversified.

The phenomenon began in the 1960s and 1970s in three areas: human rights, multinational corporations' strategies in the third world, and environmental protection. NGOs were mainly based in the English-speaking countries. For example, Amnesty International, founded in 1961 by a group of jurists, including the Irishman Sean MacBride, defines itself as "an international organization for the defense of human rights, independently of any government or ideology." The International Organization of Consumers Unions (IOCU) was created in 1960 by consumer associations in five countries (United States, Australia, Great Britain, Belgium, and the Netherlands). Over 150 organizations based in about 60 countries now belong to it. The IOCU, which has three regional offices (The Hague, Montevideo, and Penang), is at the origin of numerous campaigns, particularly against the marketing practices of agri-food and pharmaceutical companies, the pesticide industry, and, more generally, against the damaging effects of the prevailing model of development. Greenpeace was born in Vancouver in 1971 in opposition to U.S. nuclear tests and the Vietnam war. Eight years later it began its international campaigns in earnest, in favor of "green peace" and against those states or corporations that harm the environment.

As creators of the slogan "think globally, act locally," NGOs invented new modes of social intervention. While most political and trade union organizations were still reluctant to tackle the issue of the media, these organizations, with confirmed professionalism, set up their own communication systems, used the media, and succeeded in converting their causes into topical subjects. Their strength lay in their ability to organize grassroots actions and to put pressure on governmental and supranational authorities. Their networks embodied the flexibility that was sadly lacking in the large, centralized public or private organizations. They became skilled users of new technologies for collecting, storing, and analyzing information and building up networks of information exchange.

For humanitarian causes the 1980s were the years of the "charity business" and the rapid growth of media coverage, in a context marked by the downscaling of the welfare state and the erosion of public development aid policies. To collect funds, NGOs started to make intense use of persuasion methods and mailing techniques, not hesitating to use the same mailing lists as large mail-order companies. Thus, the spirit of enterprise penetrated the old humanitarian organizations and the new ones alike. The communication watchword even reached the pioneering organization of modern humanitarian aid, the Red Cross, which originated the legal concept of a neutral humanitarian space on no-man's-land ratified by the Geneva Convention in 1864. Founded in 1942, Oxfam (the Oxford Committee for Famine Relief), the ancestor of organizations for development, merged its marketing and communication departments and reorganized the 850 chain stores distributing products bought directly from third world producers by Oxfam Trading and sold under its brand name. Concepts based on development aid were definitively shelved, and a new wave of NGOs adopted names more likely to touch the public, such as Save the Children, Foster Parents Plan, and Médecins sans frontières (Doctors without Borders), better known abroad as the French Doctors.

In the 1980s, the decade par excellence of the communication myth, the growing influence of the managerial mentality generated controversy. In France debate was initiated on the excesses provoked by rallying to media logics. Objections abounded. The display by the media of poverty and its consequences and of multiple forms of violence were criticized as generating a particular view of the state of the world, and its capacity to move the public was seen as dictating the modes and places of intervention. The choice of treating emergency situations would be made at the expense of long-term action and without any concern for giving the benevolent donor a sense of responsibility. The "duty to intervene for humanitarian reasons" ignored the complexity of political solutions and provided a single-term equation based on the myth that being present on the ground was in itself a criterion of effectiveness.

However, these high-visibility debates on emergency aid overshadowed other, more discrete ruptures in relations between civil societies. Motivated by a development philosophy that challenged the terms of exchange, new forms of decentralized cooperation and networks of reciprocal exchange of knowledge appeared. The diffusionist model of persuasive communication gave way to reflection on participative democracy and the part played in it by the various media. It was in this context that, in 1983, an international association based in Quebec organized numerous united community radio stations (AMARC). This forum brought together alternative radio stations in North America, local stations in Europe, rural stations in Africa, popular stations in Latin America and Asia, as well as stations linked to specific social movements such as the women's movement. More recently, the emergence of original types of transnational groupings, such as the Euro-regions, the Mediterranean arc, the Atlantic arc, etc., spawned new noninstitutional modes of cooperation, resulting not only in multiple cultural exchanges but also in an original critical reflection about the supranational identity of peripheral regions.

Toward an International Civil Society?

A provisional evolution of NGOs' contributions to the international aggiornamento must necessarily be qualified. Their action is ambivalent. On the one hand, they serve as an alibi for governments and major financial institutions such as the World Bank, which base their projects on grassroots organizations in the fields of education and health. It is a way of ending responsibility for their international austerity policies that foment exclusion. In a sense, grassroots organizations cushion the shocks inflicted from above in the name of a financial rationality that cares little about the social costs of the measures imposed. On the other hand, the NGO model represents the stimulus that, through the conception of the everyday forms of democracy it conveys and the concrete subjects of concern it treats, reminds us of the vacuity of the abstract ideas expressed in official discourse about democracy. The organization of parallel meetings, separate from intergovernmental conferences—in 1992 in Rio, the global forum on the environment and sustainable development, and in 1995 in Peking, the counter-summit on women's issues—are examples of this, although the risk of purely symbolic action constantly threatens this new way of practicing exchange between peoples.

Through its very ambivalence, the emergence of networks of NGOs, with their diversity of origins and fields of action, constitutes a major phenomenon of the latter half of the twentieth century. These networks reflect an aspiration toward a different type of world space. Their capacity to formulate new social demands and to invent and lead new forms of struggle no longer needs to be demonstrated. Take, for example, the continuous pressure exercised by the worldwide network of NGOs on the negotiations, from 1995 to 1998, within the framework of the OECD, over the Multilateral Agreement on Investment. More than 600 organizations in nearly 70 countries, including groups as diverse as the AFL-CIO, Amnesty International, the Australian Conservation Foundation, Friends of the Earth, Oxfam, Public Citizen, the Third World Network, and the United Steelworkers of America, mobilized via an extensive website to demand the nullification of this treaty, which grew out of an outrageously excessive ideology of free trade and the deregulation of capital flows. Some geopolitical strategists have referred to this planetwide action as the harbinger of an electronically networked global civil society, but this conclusion is somewhat hasty. The spread of the organizational conception of NGOs also signifies the extension of a form of social intervention grounded in an Anglo-Saxon tradition, in phase with the prevailing empiricism and inclined to disregard the complexity of power relations in contemporary societies.

The conditions of existence of international civil society continue to depend largely on the balance of power within nation-states and the pressure exerted from these territories. Until there is proof to the contrary, and unless one adheres to the myth of the end of the state that feeds into the notion of a universal mercantile republic, the national territory remains the site where citizenship is constituted. It is the place where organized civil society, while refusing the Jacobin conception of the role of state power, can also challenge the disengagement of the state leading to complete neoliberalism. Rethinking this articulation is without doubt the best way of countering populism and its extreme nationalist forms that rest on the simplistic representation of an abstract and malevolent state, opposed to that of an idealized civil society, seen as a free space for communication between fully sovereign individuals.

The Revolts against Globalism

Throughout the world, reference to globalization has become of concern to more and more citizens. Only a mediacentric view of society can delude people into believing that a planetary perspective can be reduced to being exposed to foreign brands and transboundary information, programs, and servers. Connection with the world is also, and above all, a matter of experience. It happens through the effects of changes in the economic and social model implied by the integration of each particular society into the world space. This structural upheaval is internalized all the more by individuals because it directly affects them in their right to work, their social insurance system, and their right to benefit from public services. In the large industrial countries, more and more people draw a parallel between their situation and industrial delocation, international competition, and the aforementioned financial constraints. In other parts of the world, many countries have long since discovered the laws of the world economy through policies of austerity and the liberalization of strategic sectors of the national economy, imposed by programs for macroeconomic stabilization and structural adjustment sponsored by supranational financial institutions, as a condition for the rescheduling of their foreign debt. These reforms aim at emptying national societies of their substance by proposing new "standard" models of institutions (in education, communication, health, urban management, etc.) in keeping with globalization's logic of market fluidity. Faced with the project of a generalized economy, the paths of opposition are erratic, as the following examples of revolt indicate:

- By sporadically looting the temples of consumption or demonstrating against the structural adjustment plans of the International Monetary Fund, the rejects of globalization, who in the large cities of Latin America are surrounded by the signs of the world of consumption without having access to its products, vigorously retaliate against a capitalism that itself is assuming more and more devastating forms.
- By protesting against the decisions of urban planners to transfer libraries and daycare centers from the city center to the periphery in the name of maximizing property investments, the populations of Chengdu in the Szechwan province and other cities in southwest China expressed their concern about an export-driven free market development model focused on privileged economic areas.
- In a break with the methods of revolutionary movements in the 1960s, the neo-Zapatist guerrilla movement in the state of Chiapas in Mexico opened a new space for discussions on the future of the national identity, just as Mexico was entering the immense free market zone with the United States and Canada. This was the first armed revolt that skillfully combined a strategy of national communication, in phase with the high degree of media literacy of Mexican society, and a cross-border connection through the use of the Internet. This rebellion is all the more emblematic because Mexico was long celebrated as one of the best pupils of the World Bank—until the financial crisis of 1994 broke out, plunging the country to the limits of social collapse. Projected to the forefront of the international scene, this revolt became a symbol of resistance to the neoliberal model. It is one of the pockets of resistance referred to by Sub-Commander Marcos, i.e., the rebel groups formed by those excluded from global modernity, the "disposable people," throughout the world.
- The appearance of notions such as "cyberwar" and "netwar" in the language of Pentagon strategists shows that their conception of global security is indissociable from the new arms in the war of information. These experts speak in earnest of the great subversive potential

of the new forms of international political activism, in which they include phenomena as different as Islamic fundamentalism, the neo-Zapatist guerrilla movement in Mexico, and networks of NGOs.

- In France, persistent strikes that, in late 1995, mobilized workers from various public services (transport, post and telecommunications, gas and electricity, schools, hospitals) were interpreted as the first revolt in a G7 country against a globalization process that is remote-controlled by the financial markets. As the British historian Theodore Zeldin wrote at the time in *Le Figaro* (December 15, 1995): "What is happening in France is of universal importance. . . . The pressures at the roots of these disturbances are motivated by global causes. An era is coming to an end." In this massive labor protest movement, aimed at redefining the terms of the social contract, new processes of acquisition of a social identity within union organizations undergoing profound structural change coexisted along with outdated corporatist practices—a sign of the ambiguity of this revolt against globalism, in which the margin was necessarily narrow between withdrawal into cultural or ethnic identity and the quest for an alternative path to the universal.

The unification of the economic sphere presents a major challenge when it comes to choosing the forms of protest. It requires that social organizations, anchored in a historically situated territory but capable of broadening their scope beyond national boundaries, discover what binds them to other realities. More than 150 years after the debate between the disciples of Saint-Simon, who were partisans of the universalizing networks of industrialism, and the socialist defenders of the networks of social solidarity as the root of democratic cosmopolitanism, the question of how to build internationalism reappears, more relevant than ever, in the guise of globality. This is happening on a planet where the range of actors likely to shape it has grown but where, paradoxically, the stumbling blocks to communication and mutual understanding beyond social and cultural borders have constantly increased, as have the difficulties in recognizing the creative force of exchange between differences.

The Split between Technology and Society

Due to the acceleration of technological progress, the split between technology and society has constantly increased and gone hand in hand with increasing global asymmetry (and with the intensification of what Freud called the discontents of civilized man who has become a "prosthetic god"). No doubt, one of the most essential tasks is to reconcile citizens with a technological system with which they are currently largely unfamiliar and, as the philosopher and psychiatrist Félix Guattari advocated shortly before his death in 1992, "invent new spheres of reference so as to open the way to a reappropriation and a resymbolization of the use of communication and information tools outside of the hackneyed formulae of marketing." But it is no less essential to raise the question of citizens' appropriation, beyond individual mastery of the multimedia instrument, of the instances where the architecture of communication systems is planned. For, although it is certainly too much to expect technology to save the world, it is no less true that it constitutes a crucial element in the redefinition of the social contract and of local and national, as well as international, institutions.

We urgently need to take stock of what the philosopher Bernard Stiegler calls the "global process of externalization of memory." Systems for structuring meaning through digitization of knowledge underlie a specific geocultural model. The risk is that it may impose as a criterion of universality a particular mode of thinking and feeling, a way of "organizing the collective memory," as Simon Nora and Alain Minc put it when they diagnosed the threat of the monopolization of stocks of information by a single power. With the deployment of global cyberspace, the question is posed of knowledge being modeled by a hegemonic society that could selectively delete sections from its own collective memory. The fact that sacrosanct competitiveness was the only argument used to legitimize the historic decision taken at the G7 summit of Brussels in 1995, to rely on the market alone to favor the expansion of information highways, ought to be a serious matter of concern for informed citizens in the twenty-first century—all the more so when one knows that the organizers of this summit devoted to the global information society refused to include on the agenda the subject of "content," that is, cultural diversity, because it was seen as too controversial by nature.

Thus, gradually, with the complicity of nation-states and their supranational representatives, shielded from the indiscreet attention of civil society, the status of private enterprise as a political actor is ratified as corporate leaders demand for the "global business community" a predominant role in making the rules by which their own business is conducted.

For a Critique of Global Newspeak

> *The depths of shame were plumbed when computing, marketing, design, advertising, all the communications disciplines, seized upon the word "concept" itself, as if to say: this is our business, we are the creative people, we are conceptual! It is profoundly depressing to learn that "concept" now designates a service and computer engineering society.*

That is how Gilles Deleuze and Félix Guattari expressed, in a 1991 interview, their worry about what they called the "universals of communication" emerging, in their words, "out of puerile management-speak in its frantic dash to universalize communication so as to produce a marketable form of any concept." Distinguishing, in the history of all concepts, three "ages" (the encyclopedic age, the age of pedagogy, and the age of marketization), the two philosophers asserted that only the second age could prevent the fall from the heights of the first to the "absolute disaster for thought" represented by the third. The commodification of the universe of communication could not fail in their view to have an impact on the language of global networks. One might even say that the deregulation and privatization of communication systems goes hand in hand with a veritable deregulation of the concepts and notions we use to designate the world of networks. The influence of the technoglobalist Newspeak was only one symptom among others of the advent of social and cultural engineering that, by bringing together the different conceptual universes of corporate management operations, only deepened the ambiguity by legitimizing the return of multiple forms of empiricism. Instead of seeking to change the world, this ideology was content to describe it, without any critical examination of the vocabulary that would make understanding it possible. Thus, quite naturally, there occurs a loss of memory of the rich history of terms that, for the past two centuries, have sought to account for the conflictual quest of humanity for a postnational integration with the promise of justice and equality. This loss of memory includes the disappearance of the idea of a world-space as a plural social construction. Here is one more reason to place on the agenda a citizens' critique of the essentialist axioms of a "global village" finally coming to fruition, and a struggle against the semantic impoverishment of ways of speaking of the world and its future.

What this eschatological belief in the "information society" hides is the fact that, as the ideal of the universalism of values promoted by the great social utopias drifted into the corporate techno-utopia of globalization, the emancipatory dream of a project of world integration, characterized by the desire to abolish inequalities and injustices in the name of the imperative of social solidarity, was swept away by the cult of a project-less modernity that has submitted to a technological determinism in the guise of refounding the social bond. The ideology of limitless "communication"—but without social actors—thus takes over from the older ideology of limitless progress. This switch involves the return to a diffusionist conception of modernization associated with this ideology. In the "post-capitalist society," asserts Peter Drucker in all good faith, "the educated man of tomorrow" can expect to "live in a globalized world which will be a Westernized world." There is only a short step from such affirmations to the recycled myth of the neutrality of technology.

Conclusion

In Mutianyu, China, at the foot of a stairway leading up to the Great Wall, a granite monument was laid in 1989 with the following inscription: "Once intended to ward off enemy attacks, today it brings together the peoples of the world. The Great Wall, may it continue to act as a symbol of friendship for future generations." The emblem of the transnational corporation that financed the restoration of this section of the wall may be found just above this inscription in three languages, of which English is the most prominent.

The appropriation of this morsel of cultural heritage by an international firm says a lot about the claims of the dominant actors of the new world economy to make history and take over in the construction of the universal social bond. It attests to the ambiguous relations between the corporate world and the public sphere. Rehabilitated as social institutions, the great economic entities are constantly expanding their empire beyond the sphere of production and displaying their planetary vocation. But is the global firm able to fulfill this new mission it has conferred on itself?

"Globalization means never having to say you're sorry," said the Mexican writer and journalist Carlos Monsiváis in 1994. The global bond, a symbol of the general process of depersonalization and denationalization, empties the world of its social actors. By refusing to set any limits to their mission, by posing as managers of society as a whole, and by confiding in the self-discipline of the market as a means of governance, the world's major economic units have become irresponsible. Their "universal" ambitions mask a relentless forward march. In the late 1950s, Roland Barthes spoke of the bourgeoisie as a *société anonyme* (a limited liability company). Today, this is an appropriate name for the world business class.

Globalist arguments, the hard core of discourse on the information society, extend far beyond the question of technological networks and the circle of international firms. They embody a general way of addressing geopolitical problems and warding off all the dangers threatening the planet. Before launching the notion of a global information infrastructure, Albert Gore, in his book *Earth in the Balance,* applied the global treatment to the ecological crisis by proposing a Marshall plan for the environment, but this plan was careful not to challenge the productivist principle of the globalized economy's development model. Today, the U.S. government has a deputy Secretary of State for Global Affairs. Moreover, it is no coincidence that the notion of global human security is a recurrent feature of discourse by international authorities on the future of the world.

To allay their own uncertainties, ideologues of globalism put forward, as if they were scientific truths, the outcomes of processes whose future is anything but sure. In the name of geo-techno-financial categorical imperatives, civil societies are ordered to accept breaches of the rule of law as inexorable. Presented to citizens as accomplished facts, these decisions deprive them of their voice and push back the threshold of democratic tolerance. Taking advantage of individual and collective anxiety caused by a world that has transformed work into a privilege, the globalization of the threat of precariousness is brandished as an authoritative argument to forestall any critical outlook on dominant trends.

Throughout its historical trajectory, internationalization has been both an asset and a risk. It is more so than ever before. Fragile indeed is the project of achieving planetary cohesion by treating social change as an ancillary product of generalized economic exchange and the market mentality, and leaving it up to the monetarist approach to structure the digital society. A global solution that delegates the redevelopment of the world to Pandora is an illusion and denies citizens the right to conceive of other paths to supranational integration and a planetary consciousness capable of grasping the implications for human civilization in the current historical moment. With respect to the apparent realism professed by the universal mercantile republic, it is impossible to overstate that the only form of interdependence that warrants the quest for a grand democratic republic is an interdependence capable of freeing human communities from the obsession with unique identities and liberating mental territories from the intolerance fanned not only by exclusivist nationalisms but by free-market globalism.

"*One day everything will be fine,* that is our hope/*Everything is fine today,* that is the illusion," declared Voltaire and the thinkers of the Enlightenment. As the new century begins, the double crisis of the ideology of progress and of the great emancipatory utopias is consummated, and these utopias are replaced by that of the cybernetic prosthesis. It is fitting in these circumstances to quote Edgar Morin, who writes, "Our hope must abandon salvation. That is why I prefer to speak of tragic hope."

Selected Bibliography

Allen, R. C., ed. *To Be Continued: Soap Operas around the World.* London: Routledge, 1995.

Appadurai, A. *Modernity at Large: Cultural Dimensions of Globalization.* Minneapolis: University of Minnesota Press, 1996.

Augé, M. *Anthropology for Contemporaneous Worlds.* Stanford: Stanford University Press, 1999.

Auriac, F., and R. Brunet, eds. *Espaces, jeux et enjeux.* Paris: Fayard, 1986.

Bangemann, M. *Europe and the Global Information Society.* Brussels: European Council, 1994.

Bell, D. *The Coming of Post Industrial Society.* Harmondsworth: Penguin, 1973.

———. "Communications Technology: For Better or for Worse." *Harvard Business Review* (May–June 1979).

Bender, G., and T. Druckrey, eds. *Culture on the Brink: Ideologies of Technology.* Seattle: Bay Press, 1994.

Beniger, J. *The Control Revolution: Technological and Economic Origins of Information Society.* Cambridge: Harvard University Press, 1986.

Benjamin, W. *Paris, Capitale du XXe siècle.* Paris: Cerf, 1993.

Boltanski, L. *The Making of a Class: Cadres in French Society,* translated by A. Goldhammer. Cambridge: Cambridge University Press, 1987.

Boyd-Barrett, O. *The International News Agencies.* London: Constable, 1980.

Brady, R. A. *The Spirit and Structure of German Fascism.* London: Victor Gollancz, 1937.

Braudel, F. *Civilisation and Capitalism: 15th–18th Century,* 3 vol., translation revised by S. Reynolds. London: Collins, 1981–84.

Brown, L., ed. *State of the World 1995/1996*. Washington, D.C.: The Worldwatch Institute, 1996.

Brzezinski, Z. *Between Two Ages: America's Role in the Technetronic Era*. New York: Viking Press, 1969.

Carey, J. W. *Communication as Culture: Essays on Media and Society*. Boston: Unwin Hyman, 1989.

Castells, M. *The Power of Identity*. Oxford: Blackwell, 1997.

———. *The Rise of Network Society*. Oxford: Blackwell, 1996.

Chandler, A. D. *The Invisible Hand: The Managerial Revolution in American Business*. Cambridge: Harvard University Press, 1977.

Chesneaux, J. *World-Modernity*. London: Thames and Hudson, 1989.

Chevalier, M. "Système de la Méditerranée." *Le Globe* (February 12, 1832).

Chity, N., ed. Special issue on "Local Visions of the Global." *Journal of International Communication* 1, no. 2 (1994).

Cosgrove, D., ed. *Mappings*. London: Reaktion Books, 1999.

Crozier, M., S. Huntington, and J. Watanuki. *The Crisis of Democracy: Report on the Governability of Democracies to the Trilateral Commission*. New York: New York University Press, 1975.

Danielian, N. R. *A. T. & T.: The Story of Industrial Conquest*: NewYork: Vanguard Press, 1939.

Daugherty, W. E., and M. Janowitz, eds. *A Psychological Warfare Casebook*, published for Operations Research Office, Johns Hopkins University. Baltimore: Johns Hopkins Press, 1958.

de Certeau, M. *The Practice of Everyday Life*, vol. 1, translated by S. E. Randall. Berkeley: University of California Press, 1987.

———. *The Practice of Everyday Life*, vol. 2, translated by T. J. Tomasik. Minneapolis: University of Minnesota Press, 1998.

Deleuze, G., and F. Guattari. *Qu'est-ce que la philosophie?* Paris: Minuit, 1991.

Drucker, P. *Post-capitalist Society*. Oxford, UK: Butterworth-Heinemann, 1993.

Fanon, F. *A Study in Dying Colonialism*. New York: Monthly Review Press, 1965.

Fossaert, R. *Le Monde au 21e siècle*. Paris: Fayard, 1991.

Fourier, C. "La Fausse Industrie morcelée, répugnante, mensongère." In *Oeuvres complètes*. Reprint, Paris: Anthropos, 1966.

Freud, S. *Civilization and Its Discontents*. New York: W. W. Norton, 1961.

Fukuyama, F. "The End of History." *The Public Interest* (Summer 1989).

García Canclini, N. *Hybrid Cultures: Strategies for Entering and Leaving Modernity*, translated by C. L. Chiappari and S. L. Lopez. Minneapolis: University of Minnesota Press, 1995.

Garnham, N. *Capitalism and Global Communication: Global Culture and the Politics of Information*. London: Sage, 1990.

———. "Information Society Theory as Ideology: A Critique." *Loisir et Société/Society and Leisure* (Quebec) 12, no. 1 (1998).

Gates, B. *The Road Ahead*. New York: Viking/Penguin, 1995.

Geddes, P., and V. Brandford. *The Coming Polity*. London: Williams and Norgate, 1919.

Golding, P., and P. Harris, eds. *Beyond Cultural Imperialism: Globalization, Communication, and the New International Order*. London: Sage, 1997.

Gompert, D. C. (RAND Corporation). *Right Makes Might: Freedom and Power in the Information Age.* Headline Series, n. 316. New York: Foreign Policy Association (FPA), 1998.

Gore, A. Remarks prepared for delivery to the International Telecommunications Union Development Conference in Buenos Aires, Argentina on March 21, 1994. Released by the USIA (United States Information Agency), 1994.

Grewal, I., and C. Kaplan, eds. *Scattered Hegemonies: Postmodernity and Transnational Feminist Practices.* Minneapolis: University of Minnesota Press, 1994.

Guattari, F. *Chaosmose.* Paris: Galilée, 1992.

Guback, T. *The International Film Industry: Western Europe and America since 1945.* Bloomington: Indiana University Press, 1969.

Habermas, J. "The Public Sphere." In A. Mattelart and S. Siegelaub, eds., *Communication and Class Struggle,* vol. 1. New York: International General Editions, 1979.

Hall, E. T. *The Silent Language.* New York: Doubleday, 1959.

Hall, S., and P. du Gay, eds. *Questions of Cultural Identity.* London: Sage, 1996.

Hill, J., and Gibson P. Church, eds. *The Oxford Guide to Film Studies.* New York: Oxford University Press, 1998.

Huntington, S. "The Clash of Civilizations." *Foreign Affairs,* 72, no. 3 (Summer 1993).

———. "The Lonely Superpower." *Foreign Affairs,* vol. 78, n. 2 (1999).

King, A. D., ed. *Culture, Globalization and the World-System: Contemporary Conditions for the Representation of Identity.* Minneapolis: University of Minnesota Press, 1997.

Kropotkin, P. *Fields, Factories and Workshops, or Industry Combined with Agriculture and Brain Work with Manual Work.* London: Thomas Nelson and Sons, 1912.

Kula, W. *Measures and Men,* translated by R. Szreter. Princeton, N.J.: Princeton University Press, 1986.

Lacroix, J.G., and Tremblay G., eds. Special edition on "The 'Information Society' and Cultural Industries Theory." *Current Sociology,* 45, no. 4 (1997).

Landes, D. *The Prometheus Unbound: Technical Change and Industrial Development in Western Europe.* London: Cambridge University Press, 1969.

Lasswell, H. *Propaganda Technique in the World War.* New York: Knopf, 1927.

Lazarsfeld, P. F. "The Prognosis for International Communication Research." *Public Opinion Quarterly,* 16 (1941).

Le Bon, G. *Psychologie des foules.* English *The Crowd; A Study of the Popular Mind,* 2d ed. Dunwoody, Ga.: Norman S. Berg, 1977.

Lerner, D. *The Passing of Traditional Society: Modernizing the Middle East.* Glencoe, Ill.: Free Press, 1958.

Levitt, T. "The Globalization of Markets." *Harvard Business Review* (May–June 1983).

Liebes, T., and E. Katz. *The Export of Meaning: Cross Cultural Reading of Dallas.* Oxford: Oxford University Press, 1991.

Lippman, W. *Public Opinion.* London: Allen and Unwin, 1922.

——— and C. Merz. "A Test of the News." *The New Republic* (August 4, 1920).

Lull, J., ed. *World Families Watch Television.* Sage: London, 1988.

Lyotard, J. F. *The Postmodern Condition: A Report on Knowledge,* translated by G. Bennington and B. Massumi. Minneapolis: University of Minnesota Press, 1984.

MacBride, S., ed. *International Commission for the Study of Communication: Problems, Many Voices, One World.* Paris: UNESCO, 1980.

McChesney, R. W. *Corporate Media and the Threat to Democracy.* New York: Seven Stories Press, 1997.

McLuhan, M., and Q. Fiore. *War and Peace in the Global Village.* NewYork: Bantam House, 1968.

Mattelart, A. *Advertising International: The Privatisation of Public Space,* translated by M. Chanan. London: Routledge, 1991.

———. *The Invention of Communication,* translated by S. Emanuel. Minneapolis: University of Minnesota Press, 1996.

———. *Mapping World Communication: War, Progress, Culture,* translated by S. Emanuel and J. Cohen. Minneapolis: University of Minnesota Press, 1994.

———, and M. Mattelart. *The Carnival of Images: Brazilian Television Fiction,* translated by D. Buxton. New York: Bergin & Garvey, 1990.

———. *Theories of Communication: A Short Introduction,* translated by S. Gruenheck Taponier and J. A. Cohen. London: Sage, 1998.

Mattelart, T. *Le Cheval de Troie audiovisuel. Le rideau de fer à l'épreuve des radios et télévisions transfrontières.* Grenoble: PUG, 1995.

Mill, J. S. *Principles of Political Economy, with Some of Their Applications to Social Philosophy,* edited by W. J. Ashley. Toronto and New York: University of Toronto Press, Routledge and Kegan, 1965.

Mosco, V. *The Political Economy of Communication.* London: Sage, 1996.

Morley, D., and K. Robins. *Spaces of Identity: Global Media, Electronic Landscapes, and Cultural Boundaries.* London: Routledge, 1995.

Mumford, L., *Technics and Civilization.* New York: Harcourt, Brace & World, 1963, originally published in 1934.

Nora, S., and A. Minc. *The Computerization of Society: A Report to the President of France.* Cambridge, Mass.: MIT Press, 1980.

Nye, J. S. *Bound to Lead: The Changing Nature of American Power.* NewYork: Basic Books, 1990.

——— and W. A. Owens. "America's Information Edge," *Foreign Affairs,* 75, no. 2 (1996).

Ortiz, R. *Mundialização e cultura.* São Paulo: Brasiliense, 1994.

Paliwoda, S. J. *International Marketing.* London: Heinemann, 1986.

Palmer, M. *Des petits journaux aux grandes agences.* Paris: Aubier, 1983.

Petrella, R. "Vers un techno-apartheid global." In *Manières de voir,* quarterly edited by *Le Monde Diplomatique* (Paris), no. 18 (May 1993).

Preston, W., E. S. Herman, and H. Schiller. *Hope and Folly: The United States and UNESCO 1945–1985.* Minneapolis: University of Minnesota Press, 1989.

Raboy, M., and B. Dagenais, eds. *Media, Crisis and Democracy: Mass Communication and the Disruption of Social Order.* London: Sage, 1992.

——— et al. *Développement culturel et mondialisation de l'économie.* Quebec: Institut québecois de recherche sur la culture, 1994.

Regourd, S. *La Télévision des Européens.* Paris: La Documentation française, 1992.

Richelson, J. *A Century of Spies.* New York: Oxford University Press, 1995.

Roach, C. "Cultural Imperialism and Resistance." *Media, Culture and Society,* no. 2 (1997).

Robins, K., and F. Webster. *Times of Technoculture. From the Information Society to the Virtual Life.* London and New York: Routledge, 1999.

Rogers, E. *The Diffusion of Innovations.* Glencoe, Ill.: Free Press, 1962.

Saïd, E. *Culture and Imperialism.* New York: Alfred A. Knopf, 1993.

Saint-Simon, Henri, comte de. *Oeuvres de Saint-Simon et Enfantin.* Reprint, Paris: Anthropos, 1966.

Sauvant, K. P., and F. G. Lavipour. *Controlling Multinational Enterprises.* Boulder, Colo.: Westview Press, 1976.

Schiller, H. I. *Communication and Cultural Domination.* White Plains, N.Y.: M.E. Sharpe, 1976.

———. *Culture Inc.: The Corporate Takeover of Public Expression.* New York: Oxford University Press, 1989.

Schlesinger, P. *Media, State, and Nation: Political Violence and Collective Identities.* London: Sage, 1991.

Schmucler, H. *Memoria de la comunicación.* Buenos Aires: Biblos, 1997.

Schramm, W., and J. Riley. *The Reds Take a City: The Communist Occupation of Seoul, with Eyewitness Accounts.* New Brunswick, N.J.: Rutgers University Press, 1951.

Sinclair, J. *Peripheral Visions.* Oxford: Oxford University Press,1996.

Smith, A. *An Inquiry into the Nature and Causes of the Wealth of Nations,* edited by E. Cannan. London: Methuen, 1930.

Smythe, D. W. *Dependency Road: Communications, Capitalism, Consciousness.* Norwood, N.J.: Ablex, 1981.

Sombart, W. *Der Moderne Kapitalismus,* 3 vol. Munich and Leipzig: Dunker & Humblot, 1902–1928.

Stiegler, B. *La technique et le temps,* 2 vol. Paris: Galilée, 1994 and 1996.

Tarde, G. *L'Opinion et la foule.* Paris: Alcan, 1901.

———. *On Communication and Social Influence, Selected Papers,* edited by. T. N. Clark. Chicago: University of Chicago Press, 1969.

Veltz, P. *Mondialisation, villes et territoires.* Paris: PUF, 1996.

Virilio, P. *The Art of Motor,* translated by J. Rose. Minneapolis: University of Minnesota Press, 1995.

———. *L'Ecran du désert.* Paris: Galilée, 1991.

Wells, H. G. *Anticipations of the Reaction of Mechanical and Scientific Progress upon Human Life and Thought.* London: Chapman and Hall, 1902.

Wiener, N. *Cybernetics or Control and Communication in the Animal and the Machine.* Cambridge: 4th printing, MIT Press, 1985.

Wilden, A. *The Rules Are No Game: The Strategy of Communication.* London: Routledge & Kegan, 1987.

Williams, R. *Television: Technology and Cultural Form.* London: Fontana, 1974.

Armand Mattelart, a native of Belgium, teaches information and communication sciences at the University of Paris-VIII. He worked with the Chilean Popular Unity government from 1970 to 1973. Mattelart has authored or coauthored many books, translated into several languages, on culture, politics, and the mass media and communications theory and history. The most recent of these are *The Invention of Communication, Rethinking Media Theory: Signposts and New Directions* (with Michèle Mattelart), and *Mapping World Communication: War, Progress, Culture,* all published by the University of Minnesota Press.

Liz Carey-Libbrecht is a freelance translator. She has specialized in the translation of academic articles, books, and contributions to books in the social sciences.

James A. Cohen teaches political science at the University of Paris-VIII and the Institut d'Études Politiques of Paris. His fields of research include the political systems of contemporary Latin America and the Caribbean and problems of citizenship and ethnicity in North America and Western Europe. This is his fifth translation of a work by Armand Mattelart.